About Island Press

Since 1984, the nonprofit organization Island Press has been stimulating, shaping, and communicating ideas that are essential for solving environmental problems worldwide. With more than 1,000 titles in print and some 30 new releases each year, we are the nation's leading publisher on environmental issues. We identify innovative thinkers and emerging trends in the environmental field. We work with world-renowned experts and authors to develop cross-disciplinary solutions to environmental challenges.

Island Press designs and executes educational campaigns in conjunction with our authors to communicate their critical messages in print, in person, and online using the latest technologies, innovative programs, and the media. Our goal is to reach targeted audiences—scientists, policymakers, environmental advocates, urban planners, the media, and concerned citizens—with information that can be used to create the framework for long-term ecological health and human well-being.

Island Press gratefully acknowledges major support of our work by The Agua Fund, The Andrew W. Mellon Foundation, The Bobolink Foundation, The Curtis and Edith Munson Foundation, Forrest C. and Frances H. Lattner Foundation, The JPB Foundation, The Kresge Foundation, The Oram Foundation, Inc., The Overbrook Foundation, The S.D. Bechtel, Jr. Foundation, The Summit Charitable Foundation, Inc., and many other generous supporters.

The opinions expressed in this book are those of the author(s) and do not necessarily reflect the views of our supporters.

Prospects for Resilience

Eric W. Sanderson, William D. Solecki,
John R. Waldman, and Adam S. Parris

Prospects for Resilience: Insights from New York City's Jamaica Bay

Eric W. Sanderson, William D. Solecki,
John R. Waldman, and Adam S. Parris

ISLANDPRESS

Washington | Covelo | London

Island Press is a trademark of The Center for Resource Economics.

Keywords: Adaptive management, Biodiversity, Coastal flooding, Climate change, Combined sewer overflow, Community resilience, Computer modeling, Decision science, Disturbance, Ecological history, Estuary, Equity, Gateway National Recreation Area, Governance, Green infrastructure, Greenstreets, Hazard, Hurricane Sandy, Jamaica Bay, Monitoring, New York City, Open-source models, Panarchy, Resilience, Science and Resilience Institute at Jamaica Bay, Self-advocacy, Social-ecological system, Stakeholder input, Visionmaker model, Watershed

Library of Congress Control Number: 2016938769

✿ Printed on recycled, acid-free paper

Manufactured in the United States of America
10 9 8 7 6 5 4 3 2 1

Contents

List of Figures and Tables

FIGURES

TABLES

Acknowledgments

Like Jamaica Bay itself, this book is a testament to the powers of a resilient community.

The editors would like to thank the authors who contributed their time, energy, and wisdom to this book. Jamaica Bay's future and the practice of resilience depend on people being willing to work together and cross boundaries of discipline, institution, and experience. We appreciate their generosity, persistence, and patience as this book came together.

This book would not have happened without the efforts of Michael Dorsch. Michael was invaluable for providing technical assistance and editing of early drafts of several chapters and reviewing the text and references throughout the volume. He contributed enormously to getting this project off the ground and helping meet numerous deadlines. The editors also thank Peter Vancura for his efforts in providing technical assistance on research contributing to various chapters, including reviewing early drafts of project work and convening groups of authors. We also thank Jessice Fain, Mario Giampieri, and Chris Spagnoli for providing essential help with several of the figures in the book and Lesley Patrick, O. Douglas Price, and Sandra Clarke for logistical support.

The editors would also like to thank Bill Kornblum, John McLaughlin, and Bram Gunther for reading the entire manuscript and providing critical comments that improved many of the chapters.

The editors thank the team at Island Press. Courtney Lix, our editor, clarified our thinking and the text, provided many useful suggestions, and helped us keep the project on track. Sharis Simonian, the production editor, has been a firm hand on the tiller guiding the book to final publication. We also thank Chuck Savitt, Heather Boyer, Laurie Mazur, and David Miller at Island Press for supporting this project.

Finally, the editors wish to thank Nancy Kete and Sam Carter at the Rockefeller Foundation; the National Park Service; the City of New York; and Gillian Small, vice chancellor for research at the City University of New York, and Bill Tramontano, provost of Brooklyn College, for financial and institutional support of this project. Kate Ascher, Nerissa Moray, and Amelia Aboff of BuroHappold also provided critical support in starting the Science and Resilience Institute at Jamaica Bay. Without the support of all of these individuals, the institute and the book may never have materialized.

Eric Sanderson would like to thank John Robinson, Jon Dohlin, and Caleb McClennen at the Wildlife Conservation Society for supporting his participation. He also thanks Alison Sant and Darryl Young at the Summit Foundation for support.

PART I

Introduction to Resilience in Jamaica Bay

1

Why Prospects for Resilience for Jamaica Bay?

William D. Solecki, Eric W. Sanderson,
John R. Waldman, and Adam S. Parris

New York City woke up to issues of resilience on October 29, 2012. One might have thought that the first twelve years of the twenty-first century would have already made the point. Terrorism in 2001, major street protests in New York City in 2003 and 2011, an electricity blackout in 2003, major heat waves in 2006 and 2008, financial collapse and recession in 2009–2010, and other storms such as Tropical Storm Tammy in 2005 and Hurricane Irene in 2011, all significantly disturbed the everyday life of the city and in some cases put significant pressure on the established order. They were shocks that destroyed physical, social, and economic structures, and afterward required a process of recovery through which structures were rebuilt. The capacity of a system to recover from shocks such as these, and adapt to changing drivers and disturbances, is resilience.

What made October 29, 2012, different was the scope of the damage. That day and night Hurricane Sandy plowed into the coast of New Jersey, affecting the lives directly or indirectly of every one of the 22 million people across the New York City metropolitan region (figure 1-1). Sandy had risen as a tropical storm seven days earlier in the Caribbean, toward the end of the usual Atlantic Ocean hurricane season. The storm tracked north, gaining strength as it made landfall with hurricane-strength winds on Jamaica and then Cuba.

As the storm left land over Cuba, it seemed that it might become disorganized and dissipate, but another weather system moving west to east across North America added energy to Sandy and caused the eye of the storm to pivot toward the north-northeast, aiming at the coasts of New Jersey and New York. By the morning of October 29, Sandy was a Category 2 hurricane and had grown to a tremendous width of 1,150 miles. Although its

Figure 1-1. Hurricane Sandy, October 29, 2012. Hurricane Sandy made landfall in New Jersey on October 29, 2012, and lashed Jamaica Bay with high winds and flooding waves; the arrow points to the approximate location of Jamaica Bay in the southeast corner of New York City. The storm surge flooded many communities around Jamaica Bay and John F. Kennedy International Airport and caused a dangerous fire on Breezy Point at the tip of the Rockaway Peninsula. This book is about the prospects for resilience of urban watersheds from disturbance events, like but not limited to Sandy, for the social-ecological system (SES) that is the Jamaica Bay watershed. The image is from the Aqua-MODIS sensor. Courtesy of Jeff Schmaltz, LANCE MODIS Rapid Response Team at National Aeronautics and Space Administration's Goddard Space Flight Center.

force had diminished to tropical storm strength (approximately 75-mile-per-hour winds) when the center of Sandy landed at Brigantine, New Jersey, to the south of New York City, its effects were not only atmospheric but also oceanic. Winds drove massive waves against the shore in advance of the storm itself. When Sandy hit the coast, waves were riding on and over the top of both a storm surge and a high tide, itself amplified by the full moon and the proximity of the equinox, when high tides are at their most extreme. Storm surges that night averaged nine feet above mean sea level and in some places exceeded fourteen feet. Rain fell and fell, eventually accumulating over ten inches in some areas. Along the coasts, the deluge worsened flooding on streets already inundated by the sea.

Overnight, Sandy slowly moved inland and began to weaken, cut off from its energy supply in the warm oceanic waters. The storm had resulted in more than 150 deaths and many more injuries (Hurricane Sandy Rebuilding Task Force, 2013). Houses along the shore were blown over or lifted off their foundation, filled with sand, and soaked. In New Jersey, New York, and Connecticut, approximately 380,000 buildings were damaged or destroyed. Tens of thousands of people were displaced. Transportation, energy, food, and fuel supplies were all disrupted and would continue so for days afterward. The subways shut down because all subway tunnels were flooded by seawater, making it difficult to get to work, so many people tried to drive, causing massive traffic jams. There was a run on gasoline as rumors spread that there was not enough of it to go around given the damage to shoreline gasoline distribution facilities. New York City's mayor, Michael Bloomberg, decided to call off the New York City Marathon, leaving thousands of tourists wandering around the city in running apparel, gaping at the restaurants and stores closed for lack of power and supply. Total economic damages in the metropolitan region have been estimated at more than $60 billion, including both direct and indirect losses, making Sandy the second costliest hurricane in U.S. history (Hurricane Sandy Rebuilding Task Force, 2013) after Hurricane Katrina in 2005 (figure 1-2).

No one wants to relive the pain and anguish of Hurricane Sandy, and indeed as we write four years later, the hurricane still swirls in the consciousness of New Yorkers and their government leaders. Many parts of the city have rebuilt, some just like they were before, some in ways reflecting the experience of Sandy, and yet other places remain damaged and forlorn. With crisis comes opportunity, and one of the opportunities that Hurricane Sandy brutally unveiled was the chance to seriously think about prospects for resilience. What does it take to make a coastal city like New York resilient? How do we understand disturbances in the context of history and nature, and how do we enhance the ability of people and nature to recover after them?

These are questions that not only New York City is facing but increasingly so are other cities in the United States, such as New Orleans and Miami, and other cities around the globe, such as Amsterdam, Dhaka, and Bangkok. It is in these places that future sea level

Figure 1-2. Fire and storm damage at Breezy Point, New York. The Breezy Point fire, ignited as a result of damage from Hurricane Sandy, destroyed 130 homes and damaged many more. Breezy Point is the westernmost community on the Rockaway Peninsula, a barrier island that separates Jamaica Bay from the Atlantic Ocean (see color plate VII). The photograph was taken by Chief Mass Communication Specialist Ryan J. Courtade of the U.S. Navy, and is courtesy of the U.S. Department of Homeland Security, Federal Emergency Management Agency.

rise is forcing consideration of a series of questions about the risks and hazards of long-term use and habitation of the coastal zone.

Why Resilience of Jamaica Bay?

This book focuses on the prospects for resilience for an urban watershed where the issues come into unusual focus and clarity: Jamaica Bay (figure 1-3; see color plates). Despite the name, Jamaica Bay is not in the Caribbean, but rather it is a large coastal lagoon on the southeast side of New York City, in the boroughs of Queens and Brooklyn with a small portion in Nassau County, a neighboring New York State county. As we will learn in the second part of the book, the Jamaica Bay watershed is at the meeting place between land and sea, serves as a critical wildlife refuge, especially for migratory birds, is home to over two and a half million New Yorkers and a sight familiar to the more than 50 million air passengers peering out of windows at the twinkling green and blue waters below as their planes lift out of or land at John F. Kennedy International Airport.

Figure 1-3. A view of Jamaica Bay. Jamaica Bay and its watershed form a coupled SES where there are strong interactions between land and water, as shown in this aerial photograph of the Cross-Bay Veterans Memorial Bridge, which connects Broad Channel with the Rockaway Peninsula, shown in the foreground. The photograph was taken by Erlend Bjørtvedt in July 2012, courtesy of Wikimedia Commons.

Like many other urban estuaries, the boundaries of the bay and its environs before extensive urbanization (for Jamaica Bay, starting in the late nineteenth century) were dynamic and changeable. Shifts occurred gradually, as with century-long changes in sea level, or at punctuated moments, with disturbance events, such as the hurricanes or nor'easter storms. Although human impacts on the local ecology and landscape have long been felt, it was during the twentieth century that the edges of the bay became more hard, angular, and policy-constrained. Land reclamation, waste dumping, and large-scale infrastructure development made the bay's boundaries and jurisdictions more relevant in the everyday experience of area residents.

Jamaica Bay is a useful lens for thinking about resilience for three distinct and complementary reasons, explored in the following three sections, that we believe are also salient for other urban estuaries and watersheds. First, Jamaica Bay is a human-dominated, significantly degraded, coastal, social-ecological system (SES). The concept of SES is discussed further below, but refers to the interlinkages between the physical, ecological, and social systems of Jamaica Bay. As an interconnected SES, the people of Jamaica Bay struggle with a wide array of different disturbances operating over different time scales, from coastal storms to a complicated land use history to uneven and uncertain economic trends. Second, in large part because of its history and level of urban development, governance of Jamaica Bay has become complicated, multilayered, and in some cases, almost dysfunctional. Recently, forward-looking leaders in public and private sectors established an institute for science and resilience at Jamaica Bay as part of an agreement to co-manage the natural areas as a single ecological system. This boundary organization is critical to prospects for resilience going forward (as discussed below and in chapter 12). Third, as you will read throughout the book, relevant sciences have matured to the point where an integrated understanding of Jamaica Bay is beginning to emerge, setting the stage where we might begin to see and plan for long-term resilience.

Although the exact circumstances are particular to Jamaica Bay, they are not unique. Many parts of many urban regions throughout the United States and the world suffer from damaged and failing physical systems, ecosystems, and social systems, with the potential of disturbance from both natural and anthropogenic causes. Leaders in government and civil society are looking for models to improve all aspects of SESs. In Jamaica Bay we don't have all the answers, but we are trying. We see this book as the foundation stone of our efforts over the coming decades: to lay out what we know, to take stock of what we don't, and to summarize and synthesize the issue of resilience for ourselves and for others working on similar issues elsewhere.

Jamaica Bay: A Local Social-Ecological System with a Global Problem

Throughout the book we will refer to the concept of a social-ecological system. We believe that this concept is fundamental to resilience, not only in Jamaica Bay, but in any landscape where people and nature are in strong interaction (Folke et al., 2007). This concept is built upon observation that there are no so-called natural systems without people (at least in modern times and especially in urban areas), and, conversely, there are no social systems without nature (Berkes and Folke, 1998). From this perspective, social and ecological systems are truly interdependent and constantly coevolving and are not separate systems existing in parallel needing to be "coupled" (Tidball and Stedman, 2013). Moreover, physical, ecological, and social systems are nested. The concept of "social-ecological" emphasizes the "humans-in-the-environment perspective" (Berkes et al., 2003)—that Earth's ecosystems, from local areas to the biosphere as a whole, provide the biophysical foundation and ecosystem services for social and economic development. At the same time, the idea of an SES enables acknowledgment that ecosystems have, are, and will continue to be shaped by human actions (see figure 1-4).

From the SES perspective, a focus on only one aspect of an SES without the others is bound to lead to partial conclusions and incomplete understanding. On the one hand,

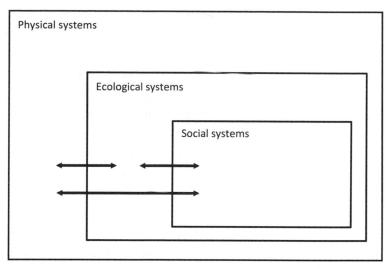

Figure 1-4. A schematic of a nested SES. Social-ecological systems can be thought of as nested physical, ecological, and social systems. Social systems are a subset of ecological systems, because human beings are organisms, and ecological systems are a subset of physical systems, which include both living and nonliving elements. The arrows indicate interactions that cross domains. Human actions influence both ecological systems and physical systems, and physical and ecological systems can interact both with and without direct human interventions. Courtesy of Eric W. Sanderson, Wildlife Conservation Society.

addressing only the biological or physical properties of a system as a basis for decision making for sustainability (a human concept) doesn't really make sense. How can human beings decide what sustainability means without considering human values and assumptions and the feedback between the two? On the other hand, some governmental and societal actors presume that only the human social domain exists, treating ecological and physical properties and processes as external and largely irrelevant to human affairs, except when "natural" disasters strike (Folke et al., 2005).

Jamaica Bay provides us with an opportunity to put the SES concept into a specific context and practice (color plate VIII). The Jamaica Bay watershed, like other estuaries near coastal cities, is a place where economic and ecological imperatives collide. For example, Jamaica Bay is the site of one of the linchpins of the global economy: John F. Kennedy International Airport. The airport projects rigidly out into the waters of the bay and nearby marshlands, constructed from sediments dredged from the bay and poured on top of the marshlands from the 1940s through the 1970s. Landfill, though, is nothing new to the bay. As in many coastal environments, once the high land was built up with homes and businesses, low-lying coastal areas were irresistible for human development and were made habitable through landfilling. Before JFK airport, Jamaica Bay was the site of New York's first municipal airport at Floyd Bennett Field, later a critical naval air station during World War II. From here, Wiley Post, Amelia Earhart, and Howard Hughes all flew out of Jamaica Bay and into history.

Jamaica Bay is also a critical part of the local economy—the gargantuan New York City metropolitan statistical area, which houses some 9 percent of Americans and provides approximately 10.3 percent of the U.S. gross domestic product (U.S. Census Bureau, 2014a; 2014b). Many commuters know the bay for the flashes of blue water visible in a sidelong glance from a speeding (or more often crawling) car, traveling the Belt Parkway, which connects Manhattan and Brooklyn with points farther east on Long Island. The Belt Parkway was a Robert Moses project to connect city people to nature by car (Caro, 1974). To build his parkway, Moses had to quash plans to create a second world-class port in Jamaica Bay, a notion bandied back and forth for three decades in the early twentieth century. Although the port was never built, the roads did enable the urbanization of the surrounding watershed. The Jamaica Bay watershed encompasses nearly half of the boroughs of Brooklyn and Queens, which had populations of nearly 2.6 million and 2.3 million, respectively, in 2014 (U.S. Census Bureau 2014a; 2014b).

With population density comes the need to deal with wastes. Here too, Jamaica Bay has served the interests of the city as if natural processes of assimilation were unlimited. For most of the twentieth century, most New Yorkers knew Jamaica Bay as one of the places where the garbage and the sewage go, with all the literal and metaphorical connotations of a dumping ground. Trash mountains were piled ignominiously into unvalued salt marshes that fringed the lagoon. Though the landfills are all now closed, their tall silhouettes stand anomalously beside the low-lying neighborhoods and coastal waters of the bay, today the tallest visible features on the coastal plain. Generations deeper in time might have remembered the dozen or so streams that once flowed down into the bay, but now nearly all the freshwater entering the bay comes from the treated effluent of four of New York City's fourteen wastewater treatment plants and two from Nassau County (NYCDEP, 2015). Though huge investments have cut back on the amount of nitrogen and other chemicals released into the bay by some 50 percent in recent years, over twenty thousand pounds of nitrogen continue to pour into the bay each day, affecting the water quality. Improvements have been made, but there is still much work to be done.

Despite the pollution, the noise, the land filling, and the history of urbanization, Jamaica Bay is also a wildlife refuge, the largest one in New York City, and a centerpiece of the Gateway National Recreation Area, an urban national park managed by the National Park Service. Prior to the National Park Service's creation of Gateway in 1971, the New York City and Nassau County Parks Departments managed the bay and today continue to manage many parks along its borders. For as long as there have been industrial developments, private community groups have focused on environmental conservation efforts, calling for enhanced protection of the bay's critical habitat and its promotion as a wildlife sanctuary. For example, famed conservationist René Dubos argued strenuously against the expansion of the JFK runways in the early 1970s. Today the American Littoral Society, Jamaica Bay Ecowatchers, New York City Audubon, The Nature Conservancy, the Natural Areas Conservancy, the Jamaica Bay-Rockaway Parks Conservancy, the Wildlife Conservation Society, and many other groups remain active in environmental education and stewardship in the area.

Because of these efforts, one can go birdwatching with a park ranger for endangered species and see the Manhattan skyline in the distance. It feels as if Jamaica Bay may have turned a corner. In 2015, 3.8 million people visited the Jamaica Bay portion of the Gateway National Recreation Area, roughly three times the number that visit Everglades National Park and roughly comparable to the number of visitors to Yellowstone and Yosemite National Parks (table 1-1). Recognition of such immense interest in urban

Table 1-1. Annual number of visits at the Jamaica Bay units of the Gateway National Recreation Area in 2015, with comparison to other national parks.

	Visitors in 2015	Visitation hours in 2015
Units at Gateway		
Canarsie Pier	75,853	227,539
Floyd Bennett Field	2,868,374	6,514,165
Fort Tilden/Breezy Point	292,818	596,187
Jamaica Bay Wildlife Refuge	178,270	429,528
Plumb Beach	22,887	68,110
Riis Park	398,160	1,221,200
Jamaica Bay units total	**3,836,362**	**9,056,729**
Gateway National Recreation Area	6,392,565	
Everglades National Park	1,077,427	
Grand Canyon National Park	5,520,736	
Yellowstone National Park	4,097,710	
Yosemite National Park	4,150,217	

Table 1-1. The national park units shown on color plate VII in and surrounding Jamaica Bay are well loved and used. In addition to these visits, it is likely that millions of visits are made to New York City parks near Jamaica Bay each year. Data compiled from the National Park Service Visitor Use Statistics (irma.nps.gov/Stats/).

nature has led to new attention by federal government agencies in parks near cities. New York City has not only the National Park Service managing parks and historic sites, but also an urban field station jointly managed by the U.S. Forest Service and the City of New York's Parks Department. The reasons people come to urban watersheds are many: people come to recreate, to fish, to lounge around on the sandy beaches, and to see the salt marshes, maritime forests, and shallow estuarine waters that provide habitat for nearly 100 species of fish, more than 325 bird species, and an unknown number of reptiles, amphibians, and mammals. On the south shore of Long Island, Jamaica Bay serves as a major stopover point along the Atlantic Flyway migration route. As an ecological area in a city, the Jamaica Bay system also provides critical seasonal or year-round support to 214 species that are on either state or federal endangered and threatened species lists (NYCDEP, 2007). Jamaica Bay is both wild and near.

Unfortunately, Jamaica Bay's prospects for resilience do not begin or end only with local influences. Sea level rise has been observed and is likely to accelerate in the twenty-first century due to global climate change, exacerbating coastal flooding (NPCC, 2015). Jamaica Bay is affected not only by the greenhouse gas (GHG) emissions from the planes, trains, and cars that move through the area on a daily basis, but also by the carbon emissions of the world. Because Jamaica Bay faces the sea at the southeast side of the city, it is critical as an intervening place where hurricanes and nor'easters threaten homes and businesses, as during Hurricane Sandy. No fewer than 62,000 single-family homes and private and public housing buildings were damaged across New York City by Hurricane Sandy, with many of those damaged by storm surge in the Jamaica Bay area (Furman Center for Real Estate and Urban Policy and Moelis Institute for Affordable Housing Policy, 2013.)

Complicated Governance of Jamaica Bay

As mentioned above, long before Sandy, Jamaica Bay had already become a complicated place to govern, which is the second critical reason why Jamaica Bay provides insights about the prospects for resilience. Some twenty-five federal, state, and local agencies have jurisdictional responsibilities in the Jamaica Bay watershed area (Table 1-2). A lot of people have a legal stake in Jamaica Bay and its watershed. In addition to hundreds of thousands of private homeowners, these public agencies have management responsibilities, as listed across the top of table 1-2. Cooperation and collaboration are obviously key, and as one might well imagine, but these have often been lacking in the past, as each agency has its own budgetary process, management structure, institutional priorities, and constituents. Further complicating management of the bay are the diversity of operational responsibilities, from water quality to environmental protection, to transportation, to military use. Each public agency, moreover, must respond to the exigencies of the local communities that surround the bay, which are diverse in terms of culture, economic status, housing type, age, and education. Everyone in New York has an opinion, and no one in New York seems shy about expressing her or his thoughts and feelings.

Yet advances are possible. Recognizing both the problems and the potential for Jamaica Bay, the City of New York and the National Park Service signed an agreement in 2012 to co-manage Jamaica Bay as an integrated SES. The City of New York manages nearly 1,000 acres of parkland and controls four wastewater treatment plants that currently provide the majority of Jamaica Bay's freshwater supply, approximately 223 million gallons per day (John McLaughlin, NYC Department of Environmental Protection, pers. comm.). The National Park Service manages the Gateway National Recreation Area, which includes approximately 9,100 acres in the Jamaica Bay watershed, including properties on the Rockaway Peninsula (NYCDEP, 2007). Under the new partnership between the City of New York and the National Park Service, the approximately 10,000 acres of federal and city-owned parks in and around Jamaica Bay would be jointly managed and initiatives created to improve ecosystem services such as recreation space, public access, and public education, while advancing research on issues related to resilience in Jamaica Bay (NYC Parks, 2015).

The agreement was signed three months before Hurricane Sandy struck. As part of this agreement, it was decided to form an institute to gather information about the science and resilience of Jamaica Bay. A group of interested scientists and researchers, community leaders, decision makers, and stakeholders were already beginning to think about how disturbances in the Jamaica Bay SES could disrupt or lead to negative changes in the region. Hurricane Sandy may have focused greater outside attention and funding on the Jamaica Bay region, but the impetus for systematically trying to assess the SES there and to begin to understand how to make the system more resilient to potential disturbances began in the years prior. Even as Sandy moved up the Atlantic seaboard, the groundwork was already being laid for the establishment of the Science and Resilience Institute at Jamaica Bay.

The Science and Resilience Institute at Jamaica Bay (SRIJB), established in 2013, aims to increase understanding of how disturbances affect natural and human systems in urban watersheds through resiliency-focused research of Jamaica Bay and to engage government and community stakeholders in the development and application of that knowledge toward a more resilient system. The institute is supported by a research consortium led by the City University of New York and is based out of its Brooklyn College campus. SRIJB represents a partnership among academic institutions, government agencies, nongovernmental organizations, and community groups. Core partnerships are sustained among the National Park Service, the City of New York, the City University of New York, Columbia University, Cornell University, the Institute of Marine and Coastal Sciences at Rutgers University, NASA Goddard Institute for Space Studies, New York Sea Grant,

Table 1-2. A partial list of public agencies with responsibility for management in the

Public Agency	Jurisdiction[2]	Air Quality	Water Quality	Wetland Management and Protection	Coastal Zone Management	Land Use Management	
U.S. Army Corps of Engineers (USACE)	F	■		■	■		
U.S. Environmental Protection Agency (EPA)	F	■		■		■	
Federal Aviation Administration (FAA)	F	■					
Fish and Wildlife Service (USFWS)	F			■			
National Park Service (NPS)	F		■	■			
U.S. Coast Guard	F		■				
NYS Department of Environmental Conservation (DEC)	S	■		■		■	
NYS Department of State	S				■	■	
NYS Office of Parks, Recreation and Historic Preservation (DPRHP)	S					■	
NYS Empire State Development Corporation	S					■	
Metropolitan Transit Authority	R					■	
NY/NJ Port Authority (NYNJPA)	R		■	■		■	
Interstate Environmental Commission	R	■	■				
NYC Department of City Planning	M				■	■	
NYC Department of Environmental Protection (DEP)	M	■	■	■			
NYC Department of Parks & Recreation (DRP)	M		■	■			
NYC Department of Transportation (DOT)	M						
NYC Department of Sanitation	M						
NYC Economic Development Corp.	M						
Borough of Brooklyn (Kings County)	C	■					
Borough of Queens (Kings County)	C	■					
Town of Hempstead	M			■	■	■	
Nassau County	C				■		

[1] This matrix, adapted from Waldman (2008), is presented as a representative rather than an exhaustive list. Other agencies may have jurisdictional responsibilities and responsibilities may exist that are not listed here. This matrix is based on input received from public and agency workshops, and not all agencies involved with management in the bay have indicated which activities they have jurisdictional responsibility for. Tinted boxes indicate areas of responsibility.

Jamaica Bay Watershed.[1]

	and Protection					Visitor Uses					Resource and Public Safety				Coexisting Uses				
	Threatened & Endangered/ Sensitive Species	Fish/Wildlife	Research–Natural Resources	Cultural and Historic Preservation	Research–Cultural Resources	Education/Interpretation	Hunting and Fishing	Water-Based Recreation	Land-Based Recreation	Concessions	Fire Management	Public Safety/Search and Rescue	Law Enforcement	Sanitation/Debris Cleanup	Dredging	Waste Management/Spills	Utilities	Transportation/Access	Military Operations

[2] Jurisdiction includes (F)ederal, (S)tate, (R)egional, (M)unicipal, (C)ounty. Other abbreviations: NYC = New York City; NYS or NY = New York State; NJ = New Jersey.

Stevens Institute of Technology, Stony Brook University (part of the State University of New York system), and the Wildlife Conservation Society.

Although the mission of SRIJB is to promote science-based policy recommendations for the bay region, coastal zone management is still managed within existing regulatory frameworks, often requiring congressionally appropriated funding. Disaster relief funds appropriated in the immediate aftermath of Sandy have provided critical resources to the Jamaica Bay region. As of 2013, $13 billion in federal funding had been appropriated to New York City for Sandy recovery efforts. Five billion dollars had been appropriated to the U.S. Army Corps of Engineers, with additional funds of $595 million from the Federal Emergency Management Agency (FEMA) and $787 million from the Department of the Interior, among other federal and state funds, to help with recovery in areas across New York City, including Jamaica Bay (NYC Recovery, 2015).

The challenge is to transition from the short-term response (and funding) to long-term resilience planning, which is driven by local, state, and national institutions, capabilities, and resources. To rebound from and prepare for future disasters, it can be difficult for public agencies to plan for and explore—with each other and the general public—regional approaches that balance human and natural systems, increase preparedness, and adapt to climate change (Parris et al., 2016). In other words, a SES approach to resilience is required that comprehends not only the science of social and ecological systems, but the politics of them.

Integrated Science and Practice Emerging

Planning for resilience requires an idea of key disturbances past and possible future and then understanding how the SES as an entirety might respond. For those ideas and understanding, we look to scientists, managers, residents, and others to provide knowledge, observations, facts, models, and statements of uncertainties. The problem is that when it comes to resilience there are many different kinds of knowledge, observations, facts, models, and uncertainties to consider, so much so that it is impossible for any one person to comprehend the task. It is only by working together in collaborative fashion that we can begin to build an integrated foundation to enhance the resilience of the SES.

Hence, this book has been created by more than fifty scholars and practitioners working together to lay the foundations to better understand what resilience in Jamaica Bay means. As described above, and throughout, our framework is that the physical, social, and ecological systems cannot be understood in isolation but only make sense as an integrated SES. Because this is not the usual way to manage either a city or a natural area, we suspect for Jamaica Bay, and for many other coastal urban estuaries, that while most of us have some of the pieces, none of us have the full picture. This book is our attempt to catalog the pieces and put them into a coherent form and message.

We are representatives of many different disciplines, including the natural sciences (oceanography, hydrology, ichthyology, ornithology, botany, conservation biology, restoration ecology, landscape and ecosystem ecology), the social sciences (geography, sociology, anthropology, planning), and the applied sciences, including engineering and decision sciences. Although as scientists we are all interested in what general lessons can be drawn about resilience in Jamaica Bay, as citizens of the region we are also interested in contributing to making Jamaica Bay more resilient, more sustainable, and overall a better place to live. We realize that while we have a voice and some important things to say, resilience does not begin or end with us; it must be an aspiration of society as a whole.

Here is the structure of the book. This chapter has laid out an introduction and established the case for why we should all care about the prospects for resilience for Jamaica Bay as an example for other places. Underscoring the need to bridge the social, biophysical, and ecological sciences to better understand and achieve resilience in complex SESs, Branco and Waldman (chapter 2), two natural scientists, provide an overview of systems thinking, complexity, and resilience theory, and then affirm the particular significance of resilience thinking in an urban estuarine watershed. The chapter sets out the conceptual framework used to engage with a resilience assessment of the Jamaica Bay SES and concludes with recommendations for managing toward resilience. Allred and colleagues (chapter 3) take up the theme of a resilience framework from a social science perspective by critically examining how people talk about Jamaica Bay through a large, systematic literature review. Chapters 2 and 3 are complementary.

Following the theoretical setup, the second part of the book explores the history and dynamics of the Jamaica Bay SES in greater detail. The section is composed of three chapters on the biophysical, ecological, and social systems. The chapters are structured to define and assess the working knowledge of the systems; how the qualities of the systems would be responsive to resilience practice; and how they relate to society's values that we hope to make resilient.

Swanson and colleagues (chapter 4) examine the biophysical systems of the bay with a particular focus on the transformation of the sediment loading and bathymetric conditions, the hydrological system and the reduction of freshwater flow into the bay, and the stresses on bay water quality and local water pollution runoff and distribution. Handel and colleagues (chapter 5) review the key historical, present, and future drivers and trends that affect Jamaica Bay's ecological systems, then explore the current resilience of a broad array of both aquatic and terrestrial flora and fauna species in the area. Current restoration projects for improved resilience of key species and ecosystems are discussed in the context of ecosystem resilience. Ramasubramanian and colleagues (chapter 6) report the results of extensive interviews in communities around Jamaica Bay to find out what resilience means for them. Reading these three chapters (4–6) in series highlights what is known and what is not with respect to our efforts to understand SES resilience.

The third section of the book turns to tools and techniques for resilience practice. Rosenzweig and colleagues (chapter 7) propose key resilience indicators essential for monitoring the resilience state of Jamaica Bay and potential impacts of future climate change. Resilience indicators are highlighted in five categories—climate hazards, water and sediment quality, biodiversity and abundance, and community. Some of these metrics can be modeled. Sanderson and colleagues (chapter 8) review the state of play of modern computational modeling approaches to issues of climate, hydrodynamics, sediments, water quality, ecological dynamics, population, transportation, and other topics important for the resilience of Jamaica Bay. Models can be used with scenarios of interventions. Catalano de Sousa and colleagues (chapter 9) explore how green infrastructure programs, encompassing natural areas as well as engineered green spaces and parks, can build resilience and reduce vulnerability to key climate risks faced in Jamaica Bay and urban coastal systems more broadly. This chapter highlights current and potential resilience infrastructure projects and also discusses the importance of adaptive planning and the involvement and education of communities in resilience infrastructure projects. Eaton and colleagues (chapter 10) round out the section on tools and techniques with a discussion relevant to Jamaica Bay stakeholders and decision makers on what decision analysis entails and what the benefits and different pathways are for harnessing decision analysis methods in resilience practice.

The book ends with chapters looking ahead to future strategies. Ramasubramanian and colleagues (chapter 11) use SES framing to suggest strategies to increase community resilience and highlight "best practices" to increase participation and inclusion. Finally, Parris and colleagues (chapter 12) discuss the new SRIJB as a boundary institution, working across the various scientific physical, ecological, and social disciplines on the one hand, and the institutional, community, and research institutions on the other. Resilience won't happen on its own, nor will it come from silos, silence, or narrow definitions. The prospects for resilience are real, but challenging, and require institutions such as the SRIJB with the bay as its focus. We also describe how lessons learned in Jamaica Bay might apply to other coastal urban ecological systems. This book represents the first major statement of the SRIJB toward a practical and scientific meaning of resilience in an urban watershed. In keeping with the adage "if you can do it in New York, you can do it anywhere," we hope that the twelve chapters in this book—the cumulative effort of more than sixty authors—will be a model for urban estuaries around the world.

References

Berkes, F., Colding, J., and Folke, C. 2003. Introduction. In: *Navigating Social-Ecological Systems: Building Resilience for Complexity and Change*. Berkes, F., Colding, J., and Folke, C. (eds). Cambridge: Cambridge University Press.

Berkes, F., and Folke, C. 1998. *Linking Social and Ecological Systems: Management Practices and Social Mechanisms for Building Resilience.* New York: Cambridge University Press.

Caro, R. 1974. *The Power Broker—Robert Moses and the Fall of New York.* New York: Knopf.

Folke, C., Hahn, T., Olsson, P., and Norberg, J. 2005. Adaptive governance of social-ecological systems. *Annual Review of Environment and Resources* 30: 441–473.

Folke, C., Pritchard, L., Berkes, F., Colding, J., and Svedin, U. 2007. The problem of fit between ecosystems and institutions: Ten years later. *Ecology and Society* 12(1): 30. Available at: http://www.ecologyandsociety.org/vol12/iss1/art30/.

Furman Center for Real Estate and Urban Policy and Moelis Institute for Affordable Housing Policy, New York University School of Law and Wagner School of Public Service. 2013. Sandy's Effects on Housing in New York City. Available at: http://furmancenter .org/files/publications/SandysEffectsOnHousingInNYC.pdf.

Hurricane Sandy Rebuilding Task Force. 2013. "Hurricane Sandy Rebuilding Strategy: Stronger Communities, A Resilient Region." Washington, D.C.: U.S. Department of Housing and Urban Development. Available at: http://portal.hud.gov/hudportal/doc uments/huddoc?id=HSRebuildingStrategy.pdf.

NPCC (New York City Panel on Climate Change). 2015. *Building the Knowledge Base for Climate Resiliency. Annals of the New York Academy of Sciences.* Available at: http://online library.wiley.com/doi/10.1111/nyas.2015.1336.issue-1/issuetoc.

NYCDEP (New York City Department of Environmental Protection). 2007. Jamaica Bay Watershed Protection Plan. Available at: http://www.nyc.gov/html/dep/html/harbor water/jamaica_bay.shtml.

NYCDEP. 2015. New York City's Wastewater Treatment System. Available at: http://www .nyc.gov/html/dep/html/wastewater/wwsystem-plants.shtml.

New York City Parks. 2015. Jamaica Bay-Rockaway Parks: A Partnership between the City of New York and the National Park Service. Available at: http://www.nycgovparks.org/ park-features/jamaica-bay-and-the-rockaways/partnership.

New York City Recovery. 2015. Sandy Funding Tracker. Available at: http://www1.nyc .gov/sandytracker/.

Parris, A.S., Garfin, B., Dow, K., Meyer, R., and Close, S.L. (eds). 2016. *Climate in Context: Science and Society Partnering for Adaptation.* Indianapolis, IN: Wiley/American Geophysical Union.

Tidball, K.G., and Stedman, R.C. 2013. Positive dependency and virtuous cycles: From resource dependence to resilience in urban social-ecological systems. *Ecological Economics* 86: 292–299.

U.S. Census Bureau. 2014a. Kings County, New York QuickFacts. Available at: http:// quickfacts.census.gov/qfd/states/36/36047.html.

U.S. Census Bureau. 2014b. Queens County, New York QuickFacts. Available at: http:// quickfacts.census.gov/qfd/states/36/36081.html.

Waldman, J. 2008. Research opportunities in the natural and social sciences at the Jamaica Bay Unit of Gateway National Recreation Area. National Park Service. Available at: http://www.nps.gov/gate/naturescience/upload/jbay-research%20opportunities.pdf.

2

Resilience Practice in
Urban Watersheds

Brett Branco and John R. Waldman

Over the past few decades, the word "resilience" has become part of the daily lexicon of scientists, resource managers, urban planners, government agencies, and members of the public who are actively involved in stewardship of natural areas. In fact, resilience has become the new paradigm for managing coastal resources and communities in an era of increasing urbanization and changing climate. For example, President Obama signed an executive order in 2013 establishing a Task Force on Climate Preparedness and Resilience. In March 2014, the mayor of New York City established the Office of Recovery and Resiliency and made resilience a goal in *One New York: The Plan for a Strong and Just City,* the 2015 update of New York City's sustainability plan.

Resilience is a word that is rich in meaning and brings together ideas and concepts that originate from an array of disciplines. Resilience is also an abstract idea that emerges from systems theory, which can seem perplexing to those who wish to embrace the resilience paradigm (Walker et al., 2012). Therefore, one of the goals of this chapter (and the larger book) is to make resilience comprehensible and specific. Here, we discuss concepts in the context of Jamaica Bay to establish a foundation of understanding upon which we can shape this new paradigm of research and management for Jamaica Bay and other urban watersheds.

The chapter is organized into four sections. First, we highlight the significance of urban estuaries and their watersheds in New York City and elsewhere, which sets the stage for why anyone should care if these estuaries and watersheds are resilient or not. Second, we outline key concepts in resilience theory and define a set of terms that reoccurs throughout the book. Third, we describe the Jamaica Bay watershed in terms of the

Walker and Salt (2012) resilience framework, as an example of the kinds of issues and questions that resilience practice presents. Fourth, and finally, we discuss next steps in advancing the theory and practice of resilience through assessment and management.

The Significance of Urban Estuarine Watersheds

A "watershed" refers to the area of land that drains into a particular body of water. An "estuary" is where freshwaters flowing from inland sources meet and mix with seawater. Estuaries are highly productive ecologically, often retaining and recycling nutrients that support abundant plankton, marsh grasses, macroalgae, invertebrates, and fishes. Jamaica Bay is not what is sometimes considered a classic estuary in which a river meets the sea, but rather a back-barrier lagoon estuary with modest total freshwater inflow from multiple but minor sources (Beck et al., 2009). Nonetheless, with its litany of anthropogenic problems, Jamaica Bay and its urbanized watershed may well serve as a model for urbanized estuarine systems throughout the world.

Because estuaries served as abundant sources of fish and shellfish to early colonists and as strategic sites for transportation between the coast and the interior, many cities are built on estuaries (as is greater New York City on the Hudson River estuary), often at their navigable limits at head of tide. Imposition of large human populations has severely degraded estuaries in many ways in the northeastern United States (Lotze, 2010). These include landfilling of shallows, dredging of channels, bulkheading, shunting and pollution of freshwater sources, direct contamination and overenrichment by industrial and human wastes and pharmaceuticals contained in sewage, colonization by nonnative species, and overfishing.

Over time (decades to centuries), the value of urban estuaries to people has shifted. Provisioning of food has declined but recreational demands have increased, often dramatically. Regular transportation on some estuaries has declined, but transportation around and across estuarine corridors is often integral to cities, given their large residential populations. Also, because many urban estuaries have improved environmentally through legislation such as the Clean Water Act of 1972, they have become more desirable as locations for residences and businesses, leading in recent times to even higher human population densities along their margins.

Thus, urban watersheds present a profound management challenge, partly because they provide two fundamentally opposed environmental services to humans—as sinks for wastes and as sources for food and recreation. Both are linked to myriad ancillary factors occurring on land and in the water, such as modifications to water flow, commercial and residential development, and transportation-based modifications. Such factors may act as "drivers" that modulate the overall resilience of an urban watershed. For instance, nitrogen overenrichment from sewage may "flip" an ecosystem into a new regime (Valiela et al., 1997). However, the entire system will also need to show resilience from large-scale external forces such as major storms. This chapter explores these concepts, both in relation to Jamaica Bay and to other urban watersheds that offer lessons in the application of notions of resilience.

Resilience: Conceptual Framework and Key Definitions

"Resilience" can be defined as the capacity for a system to tolerate or absorb disturbances as a result of either shocks or stresses to the system without shifting from one regime to another (Holling, 1973; Gunderson and Holling, 2002; Walker et al., 2004; Walker and Salt, 2006). The ability to adapt helps maintain a system's resilience. Adaptation is the process of intentional or unintentional adjustment in natural or human systems to actual or expected shocks or stresses (Walker et al., 2004; IPCC, 2014). To understand resilience, we need to understand how physical, ecological, and social systems interact.

Connections Among the Physical, Ecological, and Social Sciences

In the sciences, biophysical and ecological matters are oftentimes studied in isolation from social factors. Similarly, many social problems are also considered in a social sciences realm separate from the physical and ecological sciences. In part, these divisions stem from intellectual boundary-making characteristics of distinct disciplines in academia (Petts et al., 2008). In recent decades, however, an increasingly interdisciplinary approach is being called for (e.g., Clark et al., 2011) to explore concerns that have both social and biophysical or ecological components—for example, global environmental change issues such as climate change and resource overexploitation, sustainability, conservation, and most recently, the resilience of social-ecological systems (SES) to disturbances, as discussed in chapter 1.

The character of these challenges and the research needed to fully understand the complex dynamics of an SES require collaboration between the physical and social sciences (Berkes and Folke, 1998). Trying to understand an ecological change such as marsh loss in Jamaica Bay, for example, can no longer be considered by physical scientists alone because humans in the region have had impacts on the geomorphology of the bay as well as on water quality. To understand the human element requires that social scientists be involved as well.

Systems and Complexity

At its most basic, a system is a network of relationships among various subsidiary components, elements, or parts. Through this network, energy, matter, and information are exchanged among the system parts. An important concept for understanding the SES in Jamaica Bay and other estuaries is the recognition that the systems we are studying are both dynamic and complex. The systems are "dynamic" in that they are characterized by continual activity and change driven by internal and external forces. "Complex" here means that the functions and behavior of the system are difficult to predict, even if the functions and behavior of the individual components are relatively well known (Levin, 1998). In other words, the whole is greater than the sum of the individual parts.

Complex systems are also those that are influenced by chance events and are highly sensitive to initial conditions, such that small differences in starting conditions can translate into big differences later on. Earlier, systems theory tended to focus on linear cause and effect relationships (e.g., von Bertalanffy, 1968). However, work in recent decades has given greater attention to the idea of complexity to include nonlinearity in causal relationships, uncertainty, emergence, scale, adaptability, and self-organization of system components (e.g., Costanza et al., 1993; Holland, 1995; Levin, 1998; Berkes et al., 2003).

Complex systems may experience "adaptive cycles." An adaptive cycle is a series of phases in which the relationships among system components and flexibility in exploiting system resources changes. The four phases in an adaptive cycle are (1) rapid growth (or exploitation), (2) conservation of resources, (3) release of resources, and (4) reorganization (Holling, 2001; figure 2-1). The rapid growth phase (r phase) is when unallocated resources are plentiful and available to exploit and when innovation is rewarded and adaptive capacity is high. During the conservation phase (K phase), the system structure becomes increasingly organized and rigid, with fewer unallocated resources to exploit. During this phase, innovation and adaptation become more difficult. During the conservation phase, the system becomes vulnerable to disturbance and resilience may be low. The release phase (Ω phase) is a rapid collapse of the system structure and release of resources due to a disturbance or the crossing of a critical threshold. After a release, there is an opportunity for reorganization and renewal (α phase) during which the system's resources become up for grabs and new ideas and approaches can be adopted. The new structure that emerges may be similar to the prior structure or a new structure. For the SES of an urban estuary, the elements integral to the adaptive cycle might include physical space, financial resources, energy, and nutrients. Specific adaptive cycles within Jamaica Bay have yet to be described. However, case studies from the Tongass National Forest in Alaska (Beier et al., 2009) and the Annapurna Conservation Area in Nepal (Baral et al., 2010) provide salient examples.

Another feature of complex systems is that they can be viewed at different scales, with each connected to and influenced by scales both larger and smaller (figure 2-1). There are two types of scales to be considered. First is the physical size or geographic extent of the system to be analyzed, known as the spatial scale. Second is the time over which we wish to consider the resilience of the system, known as the temporal scale. The temporal scale

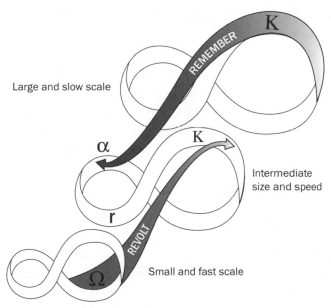

Figure 2-1. Adaptive cycles and the panarchy. The panarchy is a set of nested systems linked together in time and space. Processes at smaller and faster scales provide opportunities for innovation at larger scales, but can also trigger collapses. Meanwhile, processes at the larger scale can either stifle or promote adaptation at the intermediate scale, depending on where the larger system is in its own adaptive cycle. The image is courtesy of C.R. Allen, and C. Holling (2010). Novelty, adaptive capacity, and resilience. *Ecology and Society* 15(3): 24. [online] URL: http://www.ecologyandsociety.org/vol15/iss3/art24/.

can range from years to centuries and even beyond. This multitude of interconnected temporal and spatial scales has been described as the *panarchy* by Gunderson and Holling (2002). The main idea behind panarchy is that the experimentation and innovation that happens at the smaller scales operating on faster and shorter time scales can enable adaptation or trigger a release phase at larger scales. The latter instance is often referred to as a revolt. At the same time, the organization and structure at the larger scales provides some stability and context for the smaller scales, or memory that helps recovery in the face of disturbances. According to Walker and Salt (2012), failure to acknowledge the linkages across the scales of the panarchy is one of the common causes of resource management policy failures.

As an example, the Jamaica Bay Wildlife Refuge (JBWR) represents one scale within the panarchy of Jamaica Bay and is nested within the larger SES of the Jamaica Bay Unit of Gateway National Recreation Area (NRA), which itself is part of the National Park Service. National priorities (National Park Service scale) will provide stability and context for management and operation decisions at the Gateway NRA scale, which in turn may enable or inhibit innovative approaches to stewardship and management of the JBWR. At finer scales, the West Pond within the JBWR was breached during Hurricane Sandy, resulting in a significant increase in salinity that affected habitat critical to the maintenance of biodiversity within the JBWR. This disruption at the smaller scale within the panarchy has the potential to promote disruption and collapse at the larger scales, depending on where the systems are in the adaptive cycle. As another example, panarchy can encompass social scales from individuals to communities to the entire Jamaica Bay watershed and beyond.

Because complex systems are often nonlinear in relationships between subsystems or subsidiary components, and because there is an element of adaptability and self-organization inherent in any complex system, a more complete understanding of these systems suggests that resource management and decision making by stakeholders also must be adaptable. It is essential to recognize the need for ongoing reassessment by scientific researchers and future adaptability in the management and decision-making process.

Resilience Theory

Resilience theory concepts have emerged independently from a variety of disciplines, including psychology, engineering, and ecology. In spite of these disparate origins, there are common understandings of what resilience means. For example, "resilience" has been characterized as an emergent property of a system that can be neither predicted nor understood by examining such subsidiary components of a system (e.g., Berkes et al., 2003). Another way of thinking about emergent properties is seeing through the minutiae to the bigger picture. If one cannot look at the larger SES with a description of all the important social and ecological components expressed at the appropriate scale, it becomes difficult to understand or to predict resilience of any specific aspect of the system. For example, the water quality of Jamaica Bay can't be understood by examining nitrogen cycling in marshlands, changes in population density and land use, environmental policy instruments, and wastewater infrastructure each in isolation.

In ecology, resilience as a concept was introduced as a way to understand the confusing nonlinear dynamics being observed in natural systems. To understand resilience of a system, the concepts of regimes and regime shifts must also be understood. A regime can be thought of as a state of stability that a system can function in while maintaining *the essential identity and functions of the system*. For example, Jamaica Bay is a system characterized by extensive wetland systems that promote high biodiversity, water quality control, and flood protection, and serve as a cornerstone of the bay's aesthetic and spiritual appeal. Wetlands are essential to the system's identity and function and an aspect that we would like to be resilient to disturbances. A regime shift may occur as the result of a shock to the system as might be the case in a shore zone from an extreme natural disaster such as a tsunami or intense tropical cyclone (see chapter 1, figure 1-1). The shock may initiate a rapid transition from marsh dominance to bare mudflat, which may persist as a new stable regime (McGlathery et al., 2013). However, regime shifts may also be the result of chronic stressors that build upon one another over time and push the system toward a tipping point or threshold where it shifts from one regime or stable state to another (see chapter 9, figure 9-1). Scheffer (2009) differentiated between regime shifts associated with disturbances and chronic stressors by using the terminology "critical transitions" for the latter. The key is that when a system shifts regimes, the fundamental character, identity, and function of that system change.

For the Jamaica Bay SES, there are a multitude of possible thresholds associated with system drivers or chronic stress. For example, there could be a nitrogen loading threshold, past which the assimilation capacity of the ecosystem is exceeded and reorganization occurs. This regime shift has been described for estuaries by Valiela et al. (1997) and McGlathery et al. (2007), and it is possible that it has been surpassed in Jamaica Bay. Other thresholds that warrant consideration for the Jamaica Bay SES are population density, land use, reduction in freshwater inputs, alteration of sediment supply to the bay, total salt marsh area, and sea level. Once a threshold is exceeded, the system can rapidly transition to a new regime or state.

It is also important to recognize that a system of interest may be stuck in an undesirable regime—one that does not provide the functions that the community values. For example, as will be described in chapter 4, Jamaica Bay's wetlands are disappearing. Thus, resilience practice does not necessarily mean preserving the status quo and avoiding regime shifts. Rather, understanding the system and its resilience could allow intentional actions to overcome low-performing states and to bring the system into a new, more desirable regime. In a sense, resilience practice can be thought of as a form of intelligent adaptation that remains sensitive to existing and alternative regime states and aims to keep the system within the desirable regime.

Another way to understand the resilience concept is to consider the inputs and outputs of the SES. The components and functional relationships that make up the SES respond to external disturbances and drivers. Through the response of system variables and associated feedback loops within the system, the functions and services that constitute the system output are affected. The specification of desirable functions and services is shaped by social values and culture, as well as natural dynamics outside direct human control and, in turn, the SES functions and services can influence the culture and values of the social system, thus creating a feedback loop. More salt marsh acreage supports greater biodiversity, which attracts nature lovers who value the aesthetic, educational, and spiritual functions. This in turn fosters a culture of conservation and restoration of salt marsh and other habitat around the bay.

Qualities of Resilient Urban Systems

Although the Walker and Salt (2012) model provides one clear perspective for resilience thinking and practice that is adaptable for the Jamaica Bay SES, in moving toward implementation of resilience practice, it is important to draw from experience elsewhere. There is a rich body of literature on the characteristics of complex systems that allow them to innovate and adapt in the face of disturbances and drivers (e.g., Levin, 1998). Diversity and individuality, local interactions that are independent of a central control system, and an autonomous process for selecting good ideas are essential elements of a system that enable emergent properties, perpetuate novelty, and allow adaptation to occur. We see the elements of this thinking in recent discussions on how to build resilient cities. For example, when Ahern (2011) presented strategies for planning and designing urban resilience (multifunctionality, redundancy and modularization, multiscale networks and connectivity, and adaptive planning and design), he was really advocating for building a complex adaptive system.

The Rockefeller Foundation has recently been leading efforts to advance the resilience concept for cities. They commissioned an analysis of a city resilience framework (ARUP, 2014) using a comprehensive three-pronged approach: (1) learning from literature, (2) learning from case studies, and (3) learning from six international cities directly through fieldwork. The investigators did not favor an "asset-based" approach that focused on physical assets but rather a system-based approach that considers intangible assets such as culture, social networks, and knowledge that influence human behavior.

ARUP (2014) also found that urban systems that exhibit seven particular qualities are more likely to be resilient (table 2-1). (1) *Reflective* systems accept uncertainty and change and continuously evolve through ongoing examination of past experiences. (2) *Robust* systems anticipate potential system failures and so are designed or managed to withstand the impacts of hazard events without significant damage or loss of function. (3) *Redundant* systems incorporate spare capacity so they can accommodate disruption, extreme pressures, or surges in demand. (4) *Flexible* systems change, evolve, and adapt to changing circumstances. (5) *Resourceful* systems find means to rapidly achieve goals or needs during a shock or when under stress. (6) *Inclusive* systems emphasize broad consultation and engagement of communities, including the most vulnerable groups. (7) *Integrated* systems stress alignment between subsystems and across different scales of their operation. All of these qualities may be best promoted with a performance-based approach that defines resilience in terms of a city's ability to fulfill and sustain its core functions (ARUP, 2014).

Performance, that is, the outcome of resilience practice, offers an integrative and holistic viewpoint across the systems, assets, practices, and actions undertaken by the many actors in a typical urban setting. The core concepts of resilience are echoed in a joint report from Island Press and the Kresge Foundation (Kresge Foundation, 2015) that

Table 2-1. Attributes of a resilient city.

Category: Infrastructure and Environment

- Reduced physical exposure and vulnerability
- Continuity of critical services
- Reliable communications and mobility

Category: Health and Well-Being

- Minimal human vulnerability
- Diverse livelihoods and employment
- Adequate safeguards to human life and health

Category: Economy and Society

- Collective identity and mutual support
- Social stability and security
- Availability of financial resources and contingency funds

Category: Leadership and Strategy

- Effective leadership and management
- Empowered stakeholders
- Integrated development planning

Source: ARUP, 2014

defines urban resilience as "the capacity of a community to anticipate, plan for, and mitigate the dangers—and seize the opportunities—associated with environmental and social change." Thus, even as universal definition of resilience is elusive (and perhaps unnecessary), a consensus on the attributes of resilient urban systems, communities, and SESs is emerging.

What Is Resilience Practice?

The term resilience is no longer confined just to research of SESs but is increasingly being incorporated into resource management and stakeholder decision-making processes in ecologically sensitive areas. Inasmuch as resilience is a dynamic property of SESs, it requires a dynamic and adaptive approach. Walker and Salt (2012) advocated inclusion of three broad elements of resilience practice efforts: (1) *describing* the system, (2) *assessing* its resilience, and (3) *managing* its resilience. Here, we will use this approach to begin our discussion of resilience and resilience practice in the Jamaica Bay SES. Adopting this systematic approach facilitates the conversion of resilience theory to resilience practice and also provides a road map for others to follow as they begin to apply resilience thinking to their own urban estuaries and coastal watersheds.

Describing the Jamaica Bay Social-Ecological System in Terms of Resilience Theory

Walker and Salt (2012) recommended five steps for describing an SES for resilience assessment. These steps can be considered in any order, and it is expected that addressing them will be an iterative process. The steps are (1) the resilience *of what* (values and issues); (2) the resilience *to what* (disturbances); (3) scales (bounding the system); (4) people and governance (the players, power, and rules); and (5) drivers and trends (history and futures). Though this entire book encompasses these themes, we summarize them here for introductory purposes.

The development and implementation of a resilience-based approach to a highly urbanized estuary is a novel endeavor for which there is no simple blueprint. To begin this process for Jamaica Bay, the authors conducted a series of workshops to help launch a framework for resilience assessment for the Jamaica Bay watershed using the five steps above to guide the discussion. Our initial conclusions (and many as yet unanswered questions) are considered below. Others working in other urban coastal cities might consider conducting their own workshops at the beginning of a resilience effort.

Resilience of What?

Determining resilience *of what* in a system asks us to look at values and issues. What functions and identities of the systems need to be preserved? As with any values system, there is the potential that some components may be valued by some stakeholders and not by others. The value assigned by different individuals or groups may vary with worldviews or ideologies. Some potential tools are available to help organize and define *of what* functions and identities of the system that need to be resilient over time.

A starting point begins with the need to think of Jamaica Bay as a nested diagram of systems (i.e., human system, ecological system, physical system; see chapter 1, figure 1-4) that forms one overarching SES. Why is a healthy ecosystem important, and how does ecosystem health pair with physical barriers that serve to protect people from disturbances such as major storms? How do we extend benefits to a larger and more diverse group of people beyond the core who have regular contact with the bay? Should we extend the concept of resilience to plant and animal systems, and do they have nominal rights of their own? That is, does working within a resilience paradigm privilege one thing over another? With questions such as these that focus on values, should the answers be provided top-down from the developing institute or bottom-up from the public?

Identifying the user groups of the bay is important. It may be that the percentage of actual users in relation to the city's population as a whole is not large; however, lingering memory and emotional attachment may be greater than the current group of users. The opportunity exists to make Jamaica Bay more tangibly relevant to the whole population within its watershed and beyond it through education and direct experience. However, to do so in conjunction with resilience practice is challenging: although the simple notion of "having the property of resilience" is intuitive, resilience theory is not easily understood.

One way to look at resilience of Jamaica Bay is through consideration of its "ecosystem services," which include any positive direct or indirect benefit that wildlife or ecosystems provide to people. They include services that are being provided (e.g., recreational fishing and crabbing) and those that are not present but are desired (e.g., harvesting shellfish with safe contaminant levels). These services have changed dramatically over time (table 2-2). Originally, ecosystem services to humans emphasized provisioning, for example, finfish and shellfish for food, salt hay for livestock, flowing freshwater for drinking water. Today, any provisioning is mainly recreational, such as angling and crabbing. Moreover, a "cultural provisioning" aspect has become primary, for example, birding and kayaking.

The most crucial theme for resilience as practiced in the City of New York (SIRR, 2013) today is the protection and enhancement of human life and property. As such, the Millennium Ecosystem Assessment (United Nations Environmental Program, 2005) is a possible model inasmuch as it presents a strongly anthropocentric point of view, asking "what are the benefits that humans derive from the environment?" It can serve as a tool to help define issues and values to focus on when asking resilience *of what*? as it appraises conditions and trends in the world's ecosystems and the changing nature of their ecosystem services. At the center of the Millennium Ecosystem Assessment is human well-being, with recognition that biodiversity and ecosystems have intrinsic value as well. Thus, the functions and services that are inherent in and provided by an SES are viewed through the lens of the values that people hold.

Resilience to What?

Asking *what* the system needs to be resilient to means looking at the characteristic types of disturbances that have historically affected the system. Larger and more extreme but often infrequent disturbances may also be considered (e.g., large hurricanes). Some disturbances or shocks may be unforeseen or unknown (e.g., a new disease, such as the eelgrass wasting disease that occurred in the northeastern United States in the 1930s; Short et al., 1987). Finally, some are likely to be a result of chronic changes that gradually affect the system and may lead to significant shifts over time (e.g., population growth, land use changes, sewage discharge), though these types of disturbances may also be classified as chronic stressors. Climate change is a special case in that it will result in chronic change

Table 2-2. Selected ecosystem services provided by Jamaica Bay and its watershed.

Food (shellfish, finfish)

Genetic resources (harbors genetic diversity)

Water regulation (minimizes stormwater runoff from the watershed to the bay)

Water purification and waste treatment (denitrification is significant sink for nitrogen in the bay; wetlands intercept and mitigate contaminated runoff)

Natural hazard regulation (natural areas and built environment absorb storm surge and minimize coastal flooding impacts)

Cultural diversity

Spiritual and religious values (e.g., Hindu community, Jewish community)

Knowledge systems (perceptions of the world around us)

Educational value (e.g., opportunity to study nature in urban setting)

Aesthetic values (the beauty of nature and coastal landscapes)

Social relations (e.g., communities of people that form around shared interests in Jamaica Bay, such as birdwatching or kayaking)

Sense of place (a personal connection to Jamaica Bay)

Cultural heritage (e.g., placing high value on maintenance of historically important landscapes)

Recreation and ecotourism

Nutrient cycling (e.g., the flow and transformations of nitrogen through the ecosystem)

Primary production (creation of new organic matter through photosynthesis to support the ecosystem)

Source: Adapted from the Millennium Ecosystem Assessment (UNEP, 2005). Courtesy of the authors.

but also will exacerbate the frequency and intensity of the extreme disturbances in the Jamaica Bay watershed (Horton et al., 2015).

Disturbances can be classified as physical, biological, economic, social, or political (table 2-3). Natural physical disturbances that have affected Jamaica Bay are mostly weather events, including tropical storms such as the hurricane of 1938 and Hurricane Sandy, significant nor'easters, blizzards, and heat waves. Anthropogenic, or socially driven, physical disturbances include major infrastructure projects such as the construction of Floyd Bennett Field (opened in 1931) and JFK Airport (opened in 1948) that resulted in loss of wetlands, habitat destruction, and fragmentation. Another example is the dredging of navigation channels that changed hydrology and sediment dynamics.

Table 2-3. Selected disturbances of the Jamaica Bay social-ecological system.

Disturbance	Natural or Anthropogenic	Category
Tropical storms (e.g., hurricane of 1938, Hurricane Sandy)	Natural	Physical
Nor'easters	Natural	Physical
Blizzards	Natural	Physical
Heat waves	Natural	Physical
Construction of Floyd Bennett Field	Anthropogenic	Physical, Economic
Construction of JFK Airport	Anthropogenic	Physical, Economic
Dredging of navigation channels	Anthropogenic	Physical
Eelgrass wasting disease; MSX and Dermo outbreaks	Natural	Biological
Invasive species introduction	Natural and Anthropogenic	Biological
Great Depression in the 1930s	Anthropogenic	Economic
Recession of 2008	Anthropogenic	Economic
Neighborhood demographic changes; diaspora events	Anthropogenic	Social
Public housing construction	Anthropogenic	Social
Clean Water Act of 1972; Coastal Zone Management Act; Jamaica Bay Watershed Protection Plan	Anthropogenic	Political
Creation of Jamaica Bay Wildlife Refuge and Gateway National Recreation Area	Anthropogenic	Political

Source: Courtesy of the authors.

Table 2-3. This incomplete list shows that disturbances can be natural or anthropogenic and can be further categorized as physical, biological, economic, social or political, though most partake of multiple aspects of the social ecology of the urban watershed.

Biological disturbances can also be natural or anthropogenic. Examples include disease outbreaks that devastated local populations, such as the aforementioned eelgrass wasting disease, and shellfish diseases such as MSX (multinucleated sphere unknown) and Dermo (an intracellular parasite) that intermittently harmed oyster populations.

Invasive species are a disturbance that can have natural or anthropogenic origins. There are numerous examples in the Jamaica Bay watershed, including plants (e.g., phragmites, honeysuckle, Japanese knotweed), mammals (Norway rats, feral cats), birds (mute swans, European starlings), and marine invertebrates (Asian shore crabs, green crabs). In other cases, native species can be arguably too abundant, as with raccoons, which have been preying on diamondback terrapins (see chapter 5).

Economic, social, and political disturbances are, of course, all anthropogenic. Economic disturbances such as the Great Depression in the 1930s and the recession of 2008 can disrupt the social functions of communities around Jamaica Bay. Major infrastructure projects can cause economic disturbances by infusing local communities with rapid job and business growth. Social disturbances might include the rapid turnover of the socioeconomic demographics of neighborhoods, and could include diaspora events or the placement of large public housing initiatives within the Jamaica Bay watershed. Finally, political disturbances include shifts in policies or the creation of organizations that contribute directly to the functioning and management of the SES. Examples include the Clean Water Act of 1972, the Coastal Zone Management Act, the designation of the Jamaica Bay Wildlife Refuge and Gateway National Recreation Area, the Jamaica Bay Watershed Protection Plan, the Cooperative Management Agreement for Jamaica Bay between the National Park Service and the City of New York, and the creation of the Science and Resilience Institute of Jamaica Bay. All of these examples either established new components within the SES of Jamaica Bay or altered the functional relationship between existing components, thus serving to perturb the structure and function of the system.

Under the framework presented here, restoration projects may be viewed as intentional disturbances, as they are meant to change the components and functional relationships within the system, thus altering the ecosystem services the system provides. In Jamaica Bay, these would include habitat restoration projects at the Elders Point Islands, Spring Creek, Sunset Cove, Paerdegat Basin, Big Egg Marsh, and the landfills (Pennsylvania Avenue, Fountain Avenue, Motts Basin), as well as attempts to re-establish self-sustaining oyster populations in the bay. Some restoration projects have difficulty finding regulatory approval precisely because they change the status quo.

Disturbances, of course, differ in frequency and intensity. "Frequency" is the rate of occurrence over time, for example, the number of times a storm occurs per century. "Intensity" is the magnitude of the disturbance, such as a –2.0 on the Palmer Drought Index or the amount of precipitation that falls in twenty-four hours. The frequency vs. severity calculus is important; the system has to adapt to frequent events and must also withstand infrequent, severe events. Defining the form of resilience desired is critical—is resilience protecting or getting back to normalcy quickly? Is it better to fight flooding or to better manage it when it occurs?

Policy decisions may be as important as weather events in urban watersheds. Jamaica Bay went from a site of provisioning to one used largely for waste disposal as a result of policy change. Indeed, policy change can be seen as a kind of social disturbance—these are human actions that sometimes are not predictable, especially in terms of their long-term consequences.

At What Scale?

Defining the boundaries within which to institute resilience practice is critical. Biophysical lines are one kind of boundary; systems are always bounded geographically in some fashion. In considering resilience practice for a specified region, however, multiple systems are normally involved, such as the biophysical and the social (see chapter 1, figure 1-4). Indeed, self-organizing systems operate over multiple scales that compose the panarchy, as discussed above. Thus, two difficulties are often encountered in attempting to define scales: (1) the spatial limits of a system are not always apparent, and (2) the scales of different systems within the region often are not the same.

Boundaries of systems may be somewhat discrete (e.g., the ocean shore) or a gradient (e.g., salinity). If a gradient, it may or may not be apparent where a meaningful boundary might be recognized. Gradients may be more realistic than hard boundaries for many ecological characteristics, but at some point, value judgments must be made to assert limits. For simplicity's sake, such subjective choices may need to be single compromises among varying gradients.

Social systems also establish boundaries; however, geographic users of the bay may live beyond, and even well beyond, the watershed. Generally, biological lines rarely match with political lines, and this seems true for Jamaica Bay. An example is that transportation systems create and cut off access to the bay. The Belt Parkway that rims the northern edge of the bay is such a boundary, yet many creeks flow under it, connecting the watershed to the bay.

Maps illustrate qualities of resilience and stress and help define boundaries (color plates; also see discussion in chapter 3). These might include maps of political geography; storm surge and evacuation zones; watershed boundaries; land use; biotic communities; and keystone and endangered species. Jamaica Bay, like many other urban estuaries, has a complicated geography of jurisdictions, ecological zones, and zoning plans that need to be navigated to make the watershed more resilient.

Resilience practice must also consider multiple temporal scales. For example, a singular event may cause a severe disruption of functionality for days or weeks, such as a hypoxic episode that persists in a tributary creek. But the larger community—for example, the broader estuary—may recover its functionality sooner. The idea of persistence is important in this context, as is the appropriate time scale for measuring persistence.

For Whom and by Whom?

Defining the people and governance structures that are part of or that interact with the system is also important for building a framework for resilience assessment. Questions to be answered in defining relevant people and governance (such as the many stakeholders listed in chapter 1, table 1-2) may include the following: Who are the primary user groups? What are their rights and entitlements? Who are the secondary users? Which groups hold power? Who controls resource use and regulations at each relevant scale? Are there problems in the relationships among the controlling agencies? How do jurisdictions overlap?

And what are the relationships among the entities of different categories that govern in some way, for example, among federal, state, and local agencies, nongovernmental organizations, and community groups? What are the relative influences of entities on the bay versus those interested but based elsewhere, such as in Washington, D.C.? In Jamaica Bay, it appears that "boundary organizations" such as the Science and Resilience Institute at Jamaica Bay can provide mechanisms for coordination and collaboration, as described in chapter 11.

How Is the System Changing?

Drivers and trends are what have historically and that may in the future affect the system and cause changes. These factors might be considered in the historical profile of the system, creating the conditions under which the system exists today. Understanding these drivers and trends is important for assessing resilience because they may speak to the stability of the current system and also to how the system changed in the past and is likely to change in the future (e.g., scenario planning as a tool for describing plausible future system regimes). This historical perspective is particularly important for urban watersheds that have a long history of use, such as Jamaica Bay (detailed in chapters 4 and 5).

Drivers include ongoing changes in the global climate system, such as atmospheric composition (e.g., changes in the level of greenhouse gases), precipitation (e.g., changes in mean rainfall or the frequency of extreme events), and temperatures (e.g., changing means associated with changing climate or the frequency of extreme heat events); and human dimensions (e.g., land use management, technological changes, demographic changes, institutional arrangements, markets, and environmental and other legislative changes at federal, state, and local levels). Global change drivers may have considerable time lag and may interact with the local systems across many spatial scales.

Some of the more localized factors that mediate changes from broader (through global change) drivers include ecosystem functions, microevolutionary responses, species interactions, movement of organisms, phenology, physiology of organisms, and natural disturbances. All of these act on the makeup of the system at a given time.

Nonetheless, it also is important to consider trends in drivers in Jamaica Bay. History during the last century and a half is also something of a model and predictor for a time frame looking forward. Resilience practice should focus not just on major historical events but also on less easily perceived gradual changes.

Overall, drivers of change in Jamaica Bay include, but are not limited to, population growth; hurricanes; habitat change; use of the bay for sewage discharge, including nitrogen overenrichment; landfilling; changes to freshwater input; channel dredging; alterations of sediment sources and sinks; oyster and seagrass diseases, construction of transit infrastructure; the benefits of the Clean Water Act of 1972; zoning and planning changes; and the developing cultural understanding of the value of ecosystem services. However, it is also important to think broadly, both geographically and temporally: worldwide sea level rise and natural geomorphologic changes subsequent to the Pleistocene glaciation should be considered as important drivers of change.

Assessing the Resilience of Jamaica Bay's SES

The task of assessing resilience encompasses understanding both specified and general resilience and gaining an understanding of the system's capacity for adaptation and transformational change. Specified resilience is the resilience of some parts of the system to particular kinds of disturbances, such as those listed above, while general resilience is nonspecific. In the Walker and Salt (2012) scheme, the components described are arranged in an order that provides some insight into how the system is changing over time. And given these dynamics, we need to ask what aspects are the most important to consider.

Walker and Salt (2012) stressed the importance of assessing the overall resilience of a system, in addition to specified components. This general resilience has three important functions: (1) being able to respond quickly, (2) having reserves and access to needed resources, and (3) keeping options open. Diversity is a key aspect of general resilience. Functional diversity refers to the different functional groups in an ecosystem or different functional groups in the social domain. Response diversity is the catalog of response types, from unimodal to a broad array.

As described above, the framework for general resilience assessment facilitates identifying the resilience *of what* and *to what*. This requires looking at values and issues important to user groups and the governance regimes present in the system, as well as defining the desirable identities and functions of the system. Recognizing the significant system drivers helps build a theoretical framework for how the whole system has responded to shocks, disturbances, and adaptations in the past, and how the system may respond in the future. In other words, the framework helps us understand how the system works.

Role of Models

Identifying resilience thresholds before they are reached is difficult and provides both opportunities and challenges for the Jamaica Bay research community to provide this information to managers and stakeholders. Performing post hoc analyses of similar systems where thresholds have been crossed is one method of predicting threshold values for Jamaica Bay. Models provide a second method, though with the unfortunate reality that model validation of system resilience can be achieved only by exceeding the predicted threshold and observing the system response. Nonetheless, models are an essential component of resilience assessment and threshold prediction. A suite of models covering the full spectrum of the Jamaica Bay SES are needed, including, for example, a robust ecosystem model of Jamaica Bay (see discussion in chapter 8). An integrated modeling framework not only informs as to how components of the ecosystem interact in response to disturbance, but it also is instructive in pointing to critical missing information and, as such, helps direct future research. Obtaining basic information of this kind (e.g., understanding lower food web processes, links between nitrogen inputs and water quality, the effects of coastal engineering on flood risk) proved essential to gaining meaningful understanding in other contexts, for example, along the Hudson River, adjacent to Jamaica Bay (Suszkowski and D'Elia, 2006).

Regimes

From a management perspective, plans and actions are often directed at improving, maintaining, or transforming system structure or function. In other words, managers are implicitly or explicitly manipulating the system's identity or regime. If the Jamaica Bay SES were in a desirable regime, then the goal of resilience practice would be to perpetuate this regime. However, the current system regime may be an undesirable one, and sometimes resilience practice may be concerned with overcoming the system's resilience to transition to a more desirable regime. An example of a desirable regime might be a Jamaica Bay SES in which the local communities are not vulnerable to flooding and favorable economic and political conditions allow the time and resources to adapt to changes in flooding frequency and severity. The local community would feel connected to Jamaica Bay itself and value the ecosystem services it provides. As a result, stewardship of the bay's resources and condition would be high and there would be strong public support for restoration projects. In such a regime, positive feedback loops reinforce the improvement of the bay's resources and thus the public's perception of the bay's value.

Currently, many people around Jamaica Bay are vulnerable to coastal flooding, but the adaptations required to reduce this vulnerability are still undefined. Should we harden the shoreline (seawalls, riprap, hurricane barriers) or soften it (restore marshes, dunes, and other natural buffers)? Should we increase access and recreational opportunities in the bay or create more protected zones? Should we defend the infrastructure and property along the coast or encourage a strategic retreat? Which economic opportunities, transportation infrastructure choices, community enhancements, and governance structures are best to increase social capital? All of these considerations must be assessed not only through an understanding of the impacts on the SES structure and function, but also through the lens of our values and impacts on human well-being.

Specifying the desired system regime may be helped by referring to the attributes of resilient cities from the City Resilience Framework (ARUP, 2014) discussed earlier and to the set of ecosystem services from the Millennium Ecosystem Assessment (UNEP, 2005) (tables 2-1 and 2-2). The specification of desirable system regimes can be performed for different scales in the panarchy and in response to the questions of *resilience of what* and *resilience to what*. For example, we could define a preferred regime for the marshes of Jamaica Bay, one in which the total cumulative marsh area is enough to contribute significantly to nitrogen interception and assimilation, storm surge protection, and biodiversity. It might be possible to replace some of these functions in the face of marsh loss, for example with biological nitrogen reduction at wastewater pollution control plants, with hurricane gates, and with wave attenuators. However, loss of the island marshes would result in concurrent loss of important cultural services as well, such as aesthetic values, sense of place, and educational value. Indeed, Jamaica Bay would have an entirely different identity without its island marshes. The focal scale here then is the marshes of the bay. However, if we take a step down in the panarchy, we could narrow our focus on any one given marsh system and its immediate surroundings, such as those within Spring Creek, or the Elders Point marshes. We could also take a step up in the panarchy and focus on the mosaic of habitats throughout the bay that includes tidal flats, sandy shorelines, freshwater wetlands, and upland areas. The system identities at these different scales of the panarchy are connected but may be defined by differing sets of services and functions. Identifying these services and indeed focusing on the values held by the communities around the bay are key steps in the assessment of resilience and every bit as important as understanding the SES structure and function.

Resilience Indicator Frameworks

One potentially critical component of a resilience-based management regime is the development of key indicators that essentially "take the pulse of the system," as described more fully in chapter 5. Resilience indicators provide information on the state or level of resilience and are inextricably associated with the answers to the questions of resilience *of what* and *to what*. In other words, the set of applicable resilience indicators changes depending on the questions being addressed. The indicators can be associated with general resilience (not associated with any particular disturbance or driver) or specified resilience (associated with particular disturbances and drivers). General resilience indicators should be universal in nature and more easily applied to a range of urbanized estuaries. Specified resilience indicators may be more parochial and thus not necessarily transferable to other systems. The indicators used to assess salt marsh functionality (resilience of what?) to sea level rise and eutrophication (resilience to what?) in Jamaica Bay are examples of specified resilience indicators.

It is important to detect slowly developing disturbances or chronic stressors by monitoring some set of regular "blood pressure" measurements of the bay. These might include ecological factors such as temperature (all aspects of temperature); quantity and sources of freshwater inflow; nitrogen inputs; net primary productivity; biotic community composition and diversity; habitat mosaic; composition and distribution of marshland; and indicator species. Social metrics that could be monitored include number of visitors to a space; economic activity (value of extractive resources and services); human population size and density in watershed; total monetary value of property; recreational use; infrastructure protection; and percent impervious cover. Many of these aspects of the Jamaica Bay watershed are taken up in the chapters that follow.

Jamaica Bay: Prospects for Resilience

The assessment of resilience in Jamaica Bay's SES, like any other SES, is a continual process that must proceed in parallel with adaptive management. Although the chapters within this volume provide a snapshot of our current understanding of the components within the SES, future disturbances along with the constant change driven by chronic stressors require that assumptions, conceptual models, and predictions are revisited. The conceptual foundations provided here—complex adaptive systems, panarchy, thresholds and regime shifts, disturbances and drivers—will endure after the immediate impacts and changes initiated by any one particular disturbance, such as Hurricane Sandy, begin to fade from the memories of those who live within the boundaries of the SES. The Science and Resilience Institute at Jamaica Bay provides an opportunity to sustain the focus on resilience thinking and to assess resilience and how it changes over time using a framework that allows for learning and for transfer of new knowledge to other urban estuaries and watersheds. The prospects for resilience in Jamaica Bay are higher now than they were before the creation of the institute because of the unification of a new management framework (resilience) with a will among managers, researchers, and stakeholders to work collectively in the coproduction of knowledge about resilience in the bay.

References

Ahern, J. 2011. From fail-safe to safe-to-fail: Sustainability and resilience in the new urban world. *Landscape and Urban Planning* 100: 341–343.

ARUP. 2014. City Resilience Framework. Ove Arup & Partners International LTD, April 2014. Accessed at: https://assets.rockefellerfoundation.org/app/uploads/20150530121930/City-Resilience-Framework1.pdf.

Baral, N., Stern, M.J., and Heinen, J.T. 2010. Growth, collapse, and reorganization of the Annapurna Conservation Area, Nepal: An analysis of institutional resilience. *Ecology and Society* 15(3): 10. Accessed at: http://www.ecologyandsociety.org/vol15/iss3/art10/.

Beck, A.J., Cochran, J.K., and Sañudo-Wilhelmy, S.A. 2009. Temporal trends of dissolved trace metals in Jamaica Bay, NY: Importance of wastewater input and submarine groundwater discharge in an urban estuary. *Estuaries and Coasts* 32: 535–550.

Beier, C., Lovecraft, A.L., and Chapin, T. 2009. Growth and collapse of a resource system: An adaptive cycle of change in public lands governance and forest management in Alaska. *Ecology and Society* 14(2): 5. Accessed at: http://www.ecologyandsociety.org/vol14/iss2/art5/.

Berkes, F., Colding, J., and Folke, C. 2003. *Navigating Social–Ecological Systems: Building Resilience for Complexity and Change*. Cambridge, UK: Cambridge University Press.

Berkes, F., and Folke, C., editors. 1998. *Linking Sociological and Ecological Systems: Management Practices and Social Mechanisms for Building Resilience*. New York: Cambridge University Press.

Clark, S.G., Steen-Adams, M.M., Pfirman, S., and Wallace, R.L. 2011. Professional development of interdisciplinary environmental scholars. *Journal of Environmental Studies and Sciences* 1: 99–113.

Costanza, R., Wainger, L., Folke, C., and Mäler, K.-G. 1993. Modeling complex ecological economic systems: Toward an evolutionary, dynamic understanding of people and nature. *BioScience* 43: 545–555.

Gunderson, L.H., and Holling, C.S., editors. 2002. *Panarchy: Understanding Transformations in Human and Natural Systems*. Washington, D.C.: Island Press.

Holland, J.H. 1995. *Hidden Order: How Adaptation Builds Complexity*. Cambridge, MA: Perseus Books.

Holling, C.S. 1973. Resilience and stability of ecological systems. *Annual Review of Ecological Systems* 4: 1–23. doi:10.1146/annurev.es.04.110173.000245.

Holling, C.S. 2001. Understanding the complexity of economic, ecological and social systems. *Ecosystems* 4: 390–405. doi:10.1007/s10021-001-0101-5.

Horton, R., Bader, D., Kushnir, Y., Little, C., Blake, R., and Rosenzweig, C. 2015. Climate observations and projections. In: *New York City Panel on Climate Change 2015 Report. Annals of the New York Academy of Sciences*, 1336: 18–35. doi:10.1111/nyas.12586.

IPCC. 2014. *Climate Change 2014: Synthesis Report*. Contribution of Working Groups I, II and III to the Fifth Assessment Report of the Intergovernmental Panel on Climate Change [Core Writing Team, R.K. Pachauri and L.A. Meyer, (eds.)]. IPCC, Geneva, Switzerland, 151 pp.

Kresge Foundation. 2015. *Bounce Forward: Urban Resilience in the Era of Climate Change*. Washington, D.C.: Island Press.

Levin, S.A. 1998. Ecosystems and the biosphere as complex adaptive systems. *Ecosystems* 1(5): 431–436.

Lotze, H.K. 2010. Historical reconstruction of human-induced changes in U.S. estuaries. *Oceanography and Marine Biology: An Annual Review* 48: 267–338.

McGlathery, K.J., Reidenbach, M.A., D'Odorico, P., Fagherazzi, S., Pace, M.L., and Porter,

J.H. 2013. Nonlinear dynamics and alternative stable states in shallow coastal systems. *Oceanography* 26(3): 220–231. http://dx.doi.org/10.5670/oceanog.2013.66.

McGlathery, K.J., Sundback, K., and Anderson, I.C. 2007. Eutrophication in shallow coastal bays and lagoons: The role of plants in the coastal filter. *Marine Ecology Progress Series* 348(1): 1–18.

Petts, J., Owens, S., and Bulkeley, H. 2008. Crossing boundaries: Interdisciplinarity in the context of urban environments. *Geoforum* 39: 593–601.

Scheffer, M. 2009. *Critical Transitions in Nature and Society*. Princeton, New Jersey: Princeton University Press.

Short, F.T., Muehlstein, L.K., and Porter, D. 1987. Eelgrass wasting disease: Cause and recurrence of a marine epidemic. *Biological Bulletin* 193: 557–562.

Special Initiative for Rebuilding and Resiliency (SIRR). 2013. *PlaNYC: A Stronger, More Resilient New York*. www.nyc.gov/html/planyc2030/html/home/home.shtml.

Suszkowski, D.J., and D'Elia, C.F. 2006. The history and science of managing the Hudson River. In: *The Hudson River Estuary*, Levinton, J.S., and Waldman, J.R. (eds). 313–334. Cambridge University Press.

United Nations Environmental Program. Millennium Ecosystem Assessment. 2005. *Ecosystems and Human Well-being: A Framework for Assessment*. Washington, D.C.: Island Press.

Valiela, I., McClelland, J., Hauxwell, J., Behr, P.J., Hersh, D., and Foreman, K. 1997. Macroalgal blooms in shallow estuaries: Controls and ecophysiological and ecosystem consequences. *Limnology and Oceanography* 42(5): 1105–1118.

von Bertalanffy, L. 1968. *General System Theory: Foundations, Development, Applications*. New York: George Braziller.

Walker, B.H., Carpenter, S.R., Rockstrom, S., Crespin A.-S., and Peterson, G.D. 2012. Drivers, "slow" variables, "fast" variables, shocks, and resilience. *Ecology and Society* 17(3): 30. http://dx.doi.org/10.5751/ES-05063-170330.

Walker, B., Holling, C.S., Carpenter, S.R., and Kinzig, A. 2004. Resilience, adaptability and transformability in social–ecological systems. *Ecology and Society* 9(2): 5. Available at: http://www.ecologyandsociety.org/vol9/iss2/art5/.

Walker, B., and Salt, D. 2006. *Resilience Thinking: Sustaining Ecosystems and People in a Changing World*. Washington, D.C.: Island Press.

Walker, B.D., and Salt, D. 2012. *Resilience Practice: Building Capacity to Absorb Disturbance and Maintain Function*. Washington, D.C.: Island Press.

3

Social-Ecological System Transformation in Jamaica Bay

Shorna Allred, Bryce DuBois, Katherine Bunting-Howarth, Keith Tidball, and William D. Solecki

The view from Rulers Bar Hassock in the center of Jamaica Bay is at once wild and urban. You can watch shorebirds hunt for the eggs of horseshoe crabs, and lift your eyes to Wall Street skyscrapers on the horizon. Rulers Bar is an amalgamation of human and nonhuman processes in its own right. Having decayed in recent decades due to increasing pollution and other factors in the bay, the island is being restored through the work of the Army Corps of Engineers, the New York State Department of Environmental Conservation, the New York City Department of Environmental Protection, and the National Park Service—all aided by two groups of community activists, the Jamaica Bay Ecowatchers and the Northeast Chapter of the American Littoral Society. These community activists, and their governmental supporters, lead stewardship activities because they are concerned about the health of the bay and the health of their nearby community in Broad Channel. The Army Corps dumped sand, and more than five hundred community and youth volunteers planted more than 88,000 plugs of salt marsh cordgrass (*Spartina alterniflora*) during low tides.

Jamaica Bay can be understood as part of a network of human and nonhuman actors, where the lopsided influence of humans on the bay is becoming the dominant factor. Efforts are under way to reverse that trend. The Rulers Bar Hassock view makes the point that ecological systems of any urban estuary undergoing a resilience assessment should not be considered in a vacuum or as purely ecological, in the sense of absent from people. As discussed in chapter 2, the Jamaica Bay watershed is ultimately a social-ecological system (SES)—a multilevel or nested system that exhibits interactions to physical, ecological systems and human or social systems (Binder et al., 2013).

While chapter 2 focused on the natural science perspective on resilience, in this chapter we review the application of an SES perspective to resilience planning more generally, and in a coastal context. We then discuss how resilience frameworks such as those presented by Walker and Salt (2006, 2012) can be applied to Jamaica Bay from a social science perspective, including consideration of drivers of change, resilience of what and for whom, people and governance, and other matters related to creating SES resilience in an urban watershed.

Social-Ecological Resilience for Coastal Communities

Coasts are vital spaces of social and ecological diversity while simultaneously being sites of social and ecological vulnerability. Human populations are concentrated on the world's coasts and are at the greatest risk for future impacts from climate change and sea level rise (Jacob et al., 2007). Communities in the mid-Atlantic are exposed to hazards such as coastal flooding, hurricanes, strong winter storms, and transmission of marine-related infectious diseases (Gornitz et al., 2002). Because coasts are literal edges of ecological and cultural zones, they are sites where biological and cultural resources are mixed and shared. Thus they have the potential to be rich in biodiversity and cultural diversity, especially as sites for the sharing of multiple cultural frames (Turner et al., 2003). However, human impacts on coastal ecosystems are eroding the resilience of coastal communities. Natural barriers such as sand dunes and saltwater lagoons have protected coastal communities from minor storm surges and larger hurricane impacts; these systems, in terms of how they are managed, also rely on formal and informal institutions to respond to rapid changes (Adger et al., 2005). For example, in Jamaica Bay, restoration efforts have planted beach grass to improve barrier island habitat zones. Barrier islands include salt marshes, barrier flats, dunes, and beaches. In addition to providing valuable wildlife habitat, the salt marshes of Jamaica Bay also serve to curb shoreline erosion and provide a protective flood barrier to the neighborhoods around the bay. However, development of Jamaica Bay coastlines has resulted in loss of these important habitats, increasing communities' vulnerability to storms and their impacts. Therefore, SES resilience, and social resilience specifically, hinges on the preparation, response, and adaptation to natural disturbances in a rapidly changing environment.

The lack of community awareness of potential environmental change and disturbances contributes to vulnerability and decreases social resilience (Friesinger and Bernatchez, 2010; see chapter 6). Consequently, social learning plays an important role in preparing for and responding to coastal disturbances (see chapter 11). This was clear in the case of Hurricane Katrina (Tidball et al., 2010) and after the events of September 11, 2001 (Svendsen et al., 2014), where, in the aftermath, community members were able to work together in stewardship projects and benefited from the process of learning together. This type of deeply embedded and embodied civic ecology practice can be contrasted with the trend of rapid change in urbanization in coastal areas. The mobility of people and resources is potentially leading to the degradation of social networks and knowledge about place, and making people more vulnerable to sea level rise in places such as barrier islands on the East Coast (Bures and Kanapaux, 2011). This phenomenon is shown in the Jamaica Bay context by the different responses to Hurricane Sandy among renters and homeowners described in chapter 6. Social memory must be held and maintained both individually and institutionally, because it plays an important role in the planning, preparation, and response to disasters on the coast. Without this social memory, history is bound to repeat itself.

Social capital is one of the key drivers for adapting to change and supporting social resilience. Social capital includes social resources, networks, and connections (Aldrich, 2012). As more people in a community connect to this network, they begin to develop greater social cohesion that can be leveraged as social action toward responding and adapting to change (Adger et al., 2005; Gotham and Campanella, 2013; Colten and Giancarlo, 2011; Svendsen et al., 2014). However, social capital can erode social resilience for certain people if issues of social inequality and solidarity are not recognized in collective actions, such as in the case of post–Katrina New Orleans (Williamson, 2013).

Robust governance structures are important because they can potentially establish frameworks that prioritize social equity and diversity in adaptation strategies, which are key aspects of social resilience (Adger et al., 2005). Social equity is an important aspect of social and community resilience, because there are political and economic forces at work that can potentially push policy or actions to benefit one group or set of interests over another (Morrow, 2008). Therefore, factors that influence social resilience also include stakeholder agency and the governance process whereby ideas about resilience are generated (Larsen et al., 2011), which includes such issues as how to define the community (Breton et al., 2006). These examples show that although regime shifts may not occur when social systems go through collapse and reorganization, decisions about coastal management and development are often focused toward more rapidly changing social variables that promote economic variables as opposed to ecological variables that change more slowly. Therefore, many aspects of social resilience for coastal communities hinge on institutional support that is guided by equitable governance structures (Adger et al., 2005).

Managing for SES Coastal Resilience

Collective action toward coastal resource management is often institutionalized through management frameworks. A variety of management strategies attempt to address coastal resource management, but not all specifically engage with social or community resilience factors. The primary goal of management from a resilience perspective is to support system change and response toward a desired state, through a framework that allows for self-organization (Walker, 2006). From the particular perspective of social and community resilience, the most successful management strategies incorporate a diversity of adaptation strategies, social learning, the retention of social memory, social equity, and the support of social cohesion and collective action. Each of the following management approaches incorporates various aspects of these factors in their approaches.

Managing for resilience incorporates the need to engage in adaptive management and policy, or the need to implement "pilot" activities to assess their efficacy for the social as well as ecological aspects of the system being addressed. In some sense, it is learning by doing, or a kind of adaptive management (see discussion in chapter 11). Resilient systems management connects individuals, nonprofit organizations, government agencies, and professional or academic institutions at multiple levels; provides leadership that assists in developing trust and a vision; and provides opportunities for self-organization through social networks (Folke et al., 2005). Entities can act as bridging or "boundary" organizations by connecting groups across scales (see chapter 12). Other organizations or individuals can advance resilience through lowering the costs of collaboration by providing interventions through management approaches, providing financial assistance, proposing new laws or policies, and/or developing and implementing educational programs. Coastal management concepts have operated to support many of these goals throughout the years.

A popular SES–based governance approach is adaptive comanagement, which is a governance system involving heterogeneous actors and cross-scale interactions that involves connections that support social learning processes and encourage flexibility (Plummer, 2009). Adaptive comanagement occurs when governance structures are created to incorporate policy makers, scientists, community members, and other key informants to make decisions together (Folke et al., 2005; also see chapter 12). These adaptive governance strategies incorporate local informal managers to consider questions about resilience "for whom" and "for what end" because they are understood to be more capable of supporting recovery and have a more nuanced understanding of place (Armitage and Johnson, 2006). Olsson, Folke, and Berkes (2004) suggested that self-organization and the adaptive comanagement of ecosystems can be supported through a number of approaches. Governments can provide the social space for ecosystem management or they can provide funds for responding to environmental change and for remedial action. Efforts to support adaptive comanagement can also occur through individual or organizational support of monitoring and response to environmental feedbacks; through encouraging information flow and the development of social networks; through synthesis of various sources of disparate information, supporting lay people and policy makers to make sense of environmental change; and generally through developing arenas for collaborative learning about ecosystems and ecosystem management.

Specifically, comanagement of fisheries is practiced in many areas. In this approach, those using the resources participate in the scientific research being conducted for management purposes (Conway and Pomeroy, 2006). These research projects can include fishery stakeholder expertise, scientists, and government employees (Hartley and Robertson, 2008; Kaplan and McCay, 2004). In addition, when those who are to be regulated are allowed to participate in the regulation creation process, there is a greater chance of compliance when the regulations are implemented (Kaplan, 1998).

Another management concept that has the potential to meet many aspects of adaptive comanagement strategy is integrated coastal management (ICM) or integrated coastal zone management (ICZM). ICM is defined as "the integrated planning and management of coastal resources and environments in a manner that is based on the physical, socioeconomic and political interconnections both within and among the dynamic coastal systems, which when aggregated together, define a coastal zone" (Sorensen, 1997) or "a continuous and dynamic process by which decisions are made for the sustainable use, development, and protection of coastal and marine resources" (Cicin-Sain et al., 1998). The concept is based on the importance of the interactions between the terrestrial and marine environments and human activities, the "seamless web" that links these components together. In the 1990s, supporters of the concept highlighted the impacts that it would have on management conflicts within the coastal zone. More recently, practitioners have reflected on its ability to address modern global challenges of sustainability (Tett et al., 2013). In addition, this framework emphasizes a learning approach that requires time; a system of incentives; fostered partnerships; common agreements about knowledge, scale, and local and global learning; forging a common purpose and identity; and a diversity of communication strategies (Nursey-Bray and Harvey, 2013). Furthermore, ICM strategies must have buy-in from government institutions to prevent communities from being the first and sometimes the only responders, as is the case described in the Annapolis Basin area, Nova Scotia, Canada (Wilson and Wiber, 2009).

In addition to various management strategies, mechanisms can be put in place that help protect, enhance, and/or restore the resources within coastal areas. These institutional forces are numerous and include marine and terrestrial sanctuaries and protected areas, estuary management systems (e.g., U.S. National Estuary Program and National Estuarine Research Reserve System), special area management plans, and coastal and marine spatial planning. These kinds of institutions provide shared boundaries, intervention tools needed for resilience, and recognition of the importance of broad individual and organizational involvement to build trust and support networks in the community that support resilience.

Terrestrial ecosystems are also linked to coastal resilience through watersheds. The health of coastal systems and their ability to provide "highly valued services" is intimately linked to adjacent terrestrial systems (Millennium Ecosystem Assessment, 2005). From polluted runoff to vegetated waterways and wetland degradation, land-based management can enhance or degrade the quality of urban estuarine ecosystems (see chapter 4).

Social-Ecological Resilience in Jamaica Bay

Next, we consider how Jamaica Bay and its watershed can be framed in a resilience context, starting with drivers of change, then proceeding through a discussion of boundaries and resilience *of what* and *to what* (Walker and Salt, 2006), based on a literature review.

Drivers of Change

Drivers of change (table 3-1) vary by how often they occur and the duration of their effects and by the scale of their influence. For Jamaica Bay, short-term drivers (those with a direct influence on the SES for less than twenty-five years) with a local influence include habitat restoration and human visitor use/disturbance. Regional influences resulting from short-term drivers include the effects of 9/11 and hurricanes and strong nor'easters such as Hurricane Sandy and the hurricane of 1938.

Medium-term drivers (with a direct influence on the SES for fifty years or less) with a local influence include the development of local transportation infrastructure, such as the Belt Parkway and subway lines, Floyd Bennett Field, changes in farming and fishing, channel dredging, and practices of cultural preservation. Medium-term drivers with regional influence include John F. Kennedy International Airport; legislation passed to protect watersheds, and Jamaica Bay in particular; changes in administrative control; generational shifts from Jamaica Bay as a place of production to a place of recreation and development; and changing patterns of population.

Finally, long-term drivers influence the SES for more than fifty years. Long-term drivers with local influence include landfills (e.g., Fountain Avenue, Pennsylvania Avenue, and Edgewater landfills, all now closed), urban wastewater issues (e.g., combined sewer overflow and eutrophication), land reclamation, coastal development, and expansion of residential land use (color plate VI). Climate change is a global driver with long-term influence on Jamaica Bay. Though there may be other drivers, currently unknown, the drivers listed here signify the difficult task of managing for social-ecological resilience in an urban estuary such as the Jamaica Bay watershed.

Bounding the Jamaica Bay SES

It is important to take into account how the boundaries of any SES are being defined, and to ask whether the research questions are being investigated at a scale appropriate to garner insights in and between both the social and the ecological aspects of the SES. Walker and Salt (2006) stated that one of the initial objectives for resilience practice is to define relevant scales and boundaries of the system.

Table 3-1. Historical timeline of critical human-system interactions in the Jamaica Bay watershed.

Prior to mid-1600s	Precolonial Native Americans living in what is now New York City environs
Mid-1600s	Native Americans relinquish titles to the shore lands to the Dutch
Mid-1600s to mid-1800s	Agriculture develops on uplands; occasional mowing of salt marshes for hay; artisanal fisheries; rural communities
1812	Military blockhouse is constructed at the tip of the Rockaway Peninsula Construction of the Brooklyn Jamaica Railroad (which became part of the Long Island Railroad)
1832–1836	The Marine Pavilion (hotel) opens at Far Rockaway, Queens
1850s	Industry increases on Barren Island (Floyd Bennett Field), including first landfills
Late 1800s	Shellfish industry takes off
1898	Formation of the City of Greater New York (now New York City) incorporated Brooklyn, Queens, Jamaica, and the Rockaway Peninsula with New York City
Early 1900s	Gradual shift from place of agricultural/industrial purposes to recreation and residential communities takes place Landfills are constructed Construction of bulkheads and retaining walls becomes common
1903	Jamaica Waste Water Treatment Plant constructed (the first of four wastewater treatment plants to be built around the bay)
1905	Dredging and Jamaica Bay "improvement" begins
1910	Floyd Bennett Field created with dredged material
1917	Fort Tilden is constructed on the Rockaway Peninsula
1920s	Shellfish industry collapses
1929	Stock market collapse (beginning of the Great Depression)
1931	Floyd Bennett Field opens as first municipal airport in New York City
1934–1940s	Construction of the Belt Parkway
1935	Coney Island Wastewater Treatment Plan begins operation
1938	Jamaica Bay transferred to New York City Department of Parks and Recreation

Table 3-1. continued

1938	On September 21, the New England hurricane of 1938 makes landfall on Long Island as a Category 3 storm
Early 1940s	Idlewild Airport (what is now John F. Kennedy International Airport) construction begins
1941	Floyd Bennett Field sold to the U.S. Navy
1944	26th-Ward Wastewater Treatment Plant begins operation
1950s	Several landfills abutting the bay open
1952	Rockaway Wastewater Treatment Plant begins operation
1962	JFK runway is extended
1971	Floyd Bennett Field is deactivated by U.S. Navy
1972	Jamaica Bay transferred from New York City to National Park Service as part of Gateway National Recreation Area
1974	Fort Tilden is deactivated
1975	Eastern Airlines Flight 66 crashes on Rockaway Boulevard, Queens
1980s	Landfills close
Early 2000s	Landfill rehabilitation starts
2001	Events of September 11
	American Airlines Flight 587 crashes on the Rockaway Peninsula
2008–2009	Financial crisis and Great Recession
2012	City of New York and National Park Service sign a cooperative management agreement for Jamaica Bay
	Hurricane Sandy

Table 3-1. Understanding Jamaica Bay's resilience today requires an appreciation for how people and the ecological and physical systems have interacted over centuries.

Social science literature about Jamaica Bay has used various scales to define study areas (e.g., Black, 1981; Low, 2005; Kornblum and Van Hoorweghe, 2010). The various boundaries represent specific agency boundaries, realistic or imagined ecosystem boundaries, or boundaries drawn because of particular scientific expertise. Among the social literature about Jamaica Bay, a common boundary invoked in studies was defined by the Gateway National Recreation Area-Jamaica Bay Boundary (GNRA-JBB). The top three boundaries most frequently used in studies after GNRA-JBB include the Jamaica Bay watershed, JFK Airport, and the New York/New Jersey harbor estuary. The literature that considers Jamaica Bay from the scale of the Jamaica Bay watershed and New York/New Jersey harbor estuary is much closer to matching the framework we use in this book because it takes into account drivers and trends that exist beyond the bay front, but which still affect the basins of attraction in Jamaica Bay's SES. Other boundaries used in describing Jamaica Bay include the bordering neighborhoods; New York City; other GNRA-JBB sites (such as Fort Totten and Breezy Point); Jamaica Bay estuarine ecosystem, non-Gateway; Jamaica Bay Wildlife Refuge; pre-Gateway (NYC Parks); Jamaica Bay shoreline or waterfront; landfills bordering Jamaica Bay; and the New York Bight watershed (table 3-2).

Table 3-2. Scale of "Jamaica Bay" represented in selected studies.

Boundary Scale	Number of Studies
Gateway National Recreation Area-Jamaica Bay Unit Boundry	53
Jamaica Bay watershed	26
John F. Kennedy International Airport	22
New York/New Jersey harbor estuary	15
Bordering neighborhoods	14
New York City	14
Other Gateway National Recreation Area–Jamaica Bay Unit	13
Jamaica Bay Wildlife Refuge	10
Pre-Gateway (NYC Department of Parks and Recreation)	9
Jamaica Bay estuarine ecosystem, non-Gateway	10
Jamaica Bay shoreline or waterfront	8
Landfills bordering Jamaica Bay	8
New York Bight watershed	2

Table 3-2. People mean different geographic entities when they talk and write about Jamaica Bay. Resilience practice requires clarifying what and where people exactly mean.

People and Governance

When reviewing how people write about Jamaica Bay to gain a better understanding of its social dynamics, we found it useful to disentangle documents produced by and about the work of institutions from the complementary literature describing users of the bay. Literature about the institutions involved in work in Jamaica Bay describes a complex set of government agencies, nongovernmental organizations, scientists, visitors, residents, and activists/stewards. About twenty-five city, state, and federal agencies work or have some oversight in Jamaica Bay (see chapter 1, table 1-2). Those charged with responsibility to maintain the bay have been active in producing an extensive literature about Jamaica Bay. Several specific planning documents that drive the management of Jamaica Bay are described below.

The Jamaica Bay Watershed Protection Plan was drafted in 2007 with updates every two years, starting in 2008. The plan was drafted by the New York City Department of Environmental Protection and is the product of Local Law 71, which was signed by Mayor Bloomberg on July 20, 2005, to produce research and action toward improving the ecological health of the bay. This document laid out many recommendations, but retained the legal authority of the many distinct governmental organizations involved with the health of the bay.

Another document that is oriented specifically toward management of the bay is the new General Management Plan and Environmental Assessment from the National Park Service. This document describes a management plan that suggests providing a wide array of activities and recreational opportunities dispersed throughout the park. It includes a suggestion for increased partnerships with city and other agencies, and lays out programing and conceptual plans to increase awareness of the park.

Although there have been several attempts to establish frameworks for identifying opportunities for improving the health of the bay, there has not been an attempt to research or establish something like an integrated coastal management program. Furthermore, these documents have not incorporated an SES perspective, thus leading to neglected social and/or ecological drivers, depending on the focus of the document.

In addition to documents that describe the role of managers and government agencies in Jamaica Bay, we identified several bodies of work that touch on the users (table 3-3) of the bay (Kornblum, 1983; Burger, 2000; Kurlansky, 2007). This literature describes many different types of recreational visitors (for example, birders, fishermen, boaters, students and children, and bicyclists), in addition to other visitors, residents, and stewards/activists involved in sociopolitical activities to pursue various desired goals. In some cases, stewards were involved in the drafting of research documents, while in others they were described as ethnographic key informants. Missing in this literature are descriptions of environmental justice issues and the sociopolitical struggles involved in pursuing justice in these places.

Resilience of What?

In answering the question resilience *of what*, using the research about Jamaica Bay reviewed here, the Jamaica Bay estuarine ecosystem, as well as human life and property, figure prominently. Many studies have focused on human impacts that influence

Table 3-3. People and institutions of Jamaica Bay represented in selected studies.

People and Institutions	Number of Studies
People	
Visitors, unspecified	29
Active recreationists	18
Residents	13
Passive recreationists	2
Stewards/activists	4
Scientists/institutes	41
Nongovernmental organizations	7
Government agencies with responsibility to maintain Jamaica Bay	
MONITORING (U.S. Army Corps of Engineers, New York State Department of Environmental Conservation, NYC Department of Environmental Protection, National Park Service, Natural Resources Group: NYC Parks, U.S. Environmental Protection Agency, U.S. Geological Survey)	43
RESTORATION (NYC Department of Environmental Protection, New York State Department of Environmental Conservation, National Park Service, U.S. Fish and Wildlife Service)	15
HOLISTIC (NYC Department of City Planning)	6
TRANSPORTATION/OTHER (Port Authority of New York and New Jersey, JFK International Airport, National Oceanic and Atmospheric Administration)	19

Table 3-3. A wide range of institutional actors are at work in Jamaica Bay. Such concerted attention lays a groundwork for resilience but also complicates management practice and action. Meanwhile, community concerns are not always heard or acted upon.

the overall health of the bay, particularly nitrogen released by the wastewater treatment plants (e.g., Benotti et al., 2007) (table 3-4). Research on human property and human life focuses on development around the bay, cultural resources that are found within the GNRA-JBB, visitor uses, and significant sociocultural relationships with the bay, such as fishing, boating, and beach going. Other less frequent citations include research on the human impacts on birds (for example, colonial species, herons, egrets, shorebirds, piping plovers, and Canada geese), horseshoe crabs, marshes within Jamaica Bay, diamondback terrapins, and fish, shellfish, and bivalves.

Table 3-4. Targets of resilience efforts as represented in selected studies.

Resilience of What?	Number of Studies
Jamaica Bay estuarine ecosystem	72
Human property/life	70
Birds	25
John F. Kennedy International Airport	19
Marshes	15
General biodiversity	2
Fish, shellfish, and bivalves	5
Piping plover	4
Horseshoe crab	2
Diamondback terrapin	2
Other	1

Table 3-4. Different people and institutions have different values for the Jamaica Bay watershed, which are reflected in this summary of what aspects of the SES might be made more resilient in the future.

None of this literature employs an SES perspective, per se, though there is recognition of linked relationships between humans and the Jamaica Bay watershed in much of research. Not explicitly linking the social to the ecological suggests that there is still need for research that works to identify key drivers of change and how the ecosystem and the people within it are affected.

Resilience to What?

The social literature about Jamaica Bay contains many descriptions of disturbances (table 3-5). From an SES resilience perspective, these can be lumped into two general categories; characteristic and large infrequent disturbances. Characteristic disturbances are those that you know and expect. Large infrequent disturbances are often similar in type to characteristic disturbances but are rarer and significantly larger in magnitude (Walker and Salt, 2006).

The social literature about Jamaica Bay describes several types of known and expected "characteristic" disturbances. For example, severe storms and sea level rise associated with climate change are known and expected (Gornitz et al., 2002; Hartig et al., 2002; Jacob et al., 2007). Sea level rise will affect the Jamaica Bay estuarine ecosystem in numerous ways, including loss of habitat for horseshoe crabs (Anthony et al., 2009), increased inundation

Table 3-5. Characteristic disturbances as identified in the published Jamaica Bay literature.

Types of Characteristic Disturbances	Number of Studies
Human Visitors, Use, and Recreation	
Visitor use (for example, fishing)	29
Impacts on piping plover, gulls, and diamond-back terrapin	11
Shoreline Development, Hard Structures, and Dredging	
Increased tidal ranges (horseshoe crabs; marshes)	13
Other	20
JFK International Airport	
Bird strikes, noise, loss of nesting habitat (gulls, Canada geese, and others), gull reduction; people (air pollution)	20
Nitrogen loading and development (marsh), nitrogen, hydrocarbon release, marsh loss, development (JB estuarine ecosystem)	4

Table 3-5. Similar to the diversity of targets of resilience, the kinds of disturbances relevant to Jamaica Bay are varied.

of marshes (Hartig et al., 2000, 2002), and increased flooding, especially associated with severe storms such as Hurricane Sandy (Jacob et al., 2007).

Human use and recreation within the Jamaica Bay ecosystem has been described as a disturbance affecting piping plovers, laughing gulls, and diamondback terrapins, while fishing has been monitored for human health concerns (e.g., Burger, 2000; Goldin, 1993; Waldman, 2008). Over time, shoreline development, the building of hard shoreline structures, and the dredging of Jamaica Bay have increased tidal ranges, affecting horseshoe crabs and marshes (Hartig et al., 2002). Finally, JFK Airport poses a characteristic disturbance for many aspects of the Jamaica Bay watershed at the same time that it provides important economic benefits for the city and region. Birds, for example, are affected by airplane noise and bird strikes, the loss of nesting habitat due to runway extensions (for birds such as gulls, Canada geese, and others), and gull reduction programs to reduce bird strikes (Buckley and McCarthy, 1994; Burger, 1983, 1985).

For the estuarine ecosystem in general and marshes in particular, nitrogen loading from residential and commercial runoff and development of the shoreline around the airport have been found to have significant impacts. More general characteristic disturbances from JFK include air pollution and hydrocarbon release (however, only human impacts from JFK air pollution have been studied).

Large infrequent disturbances described in the social literature about Jamaica Bay include hurricanes (leading to flooding impacts and other disruptions and releases from combined sewer overflows; Kenward et al., 2013), the impacts of the terrorist events on 9/11 (neighborhood impacts and pollution; Boyle, 2002; Hildebrandt, 2005), and climate change and sea level rise. Wastewater and landfill impacts constitute the largest body of literature on infrequent disturbance—approximately 80 percent of all references identified. These concern a long list of pollutants such as leachates, dissolved trace/toxic metals, pharmaceuticals, nitrogen loading, coliform bacteria, eutrophication affecting the Jamaica Bay estuarine ecosystem, polychlorinated biphenyls (PCBs), organochlorine pesticides and mercury affecting marshes, and more general public health concerns relating to people.

One recent series of scientific inquiries has attempted to document and identify the causes of marsh loss (e.g., Hartig et al., 2002; Gateway National Recreation Area, 2007; Swanson and Wilson, 2008; Wigand et al., 2014), which have been attributed to a variety of different factors, including sea level rise, pollution, wave action, sediment supply changes, and so on. It will be important to continue to pursue research that identifies characteristic, large infrequent, and as yet unknown, disturbances as we move into a time when climate change will continue to change the frequency and types of disturbances in the Jamaica Bay SES.

Key Gaps in SES Research on Jamaica Bay

We conclude by noting some gaps in the current understanding of Jamaica Bay as an SES. The gaps range from the lack of social-ecological research at the watershed level to research that addresses the needs of vulnerable populations. Understanding and addressing knowledge gaps, in any SES being studied, is a key element of being able to understand the complex facets of resilience.

SES Research at the Watershed Level

Although there are examples of social science literature that investigated social system impacts and characteristics in the Jamaica Bay watershed, most literature has focused on specific neighborhoods. It's rarer to find investigations at multiple scales related to the Jamaica Bay watershed. One example to highlight as a step in the right direction is Kornblum and Van Hooreweghe (2010). This Jamaica Bay Ethnographic Overview and Assessment was prepared as part of the National Park Service's Ethnography Program to "assist managers and planners of Gateway National Recreation Area to better understand changes in uses of Jamaica Bay resources since the National Recreation Area was created by Congress in 1972" (p. iii). The report traces historical settlement patterns and socio-demographic characteristics, and includes descriptions of various cultural meanings and attachments to the bay and watershed. The result is a document, and later dissertation with related findings (Van Hooreweghe, 2012), that helps planners and managers understand the relationships between people and place; however, these documents do not pursue challenges and solutions related to the resilience of the bay as an SES. Additionally, social science research on the bordering neighborhoods and the sociocultural issues exists, but it is far outnumbered by research focused on the bay and nearby estuarine ecosystems. This finding is of concern because it suggests that much research has not also considered social aspects from the watershed level. This may potentially limit the ability of managers and policy makers to connect the appropriate action or response to a social issue through scale matching.

Vulnerable Populations

In our case of the Jamaica Bay watershed, there are few accounts of the environmental justice issues and the sociopolitical struggles involved in pursuing justice based on an SES perspective. Therefore, we identified a need for research with vulnerable populations to identify and pursue research questions that were relevant to them. Sociological literature helped to guide us to this conclusion through socio-historical accounts of isolated residents living in public housing in Rockaway (Kaplan and Kaplan, 2003), concerns about unequal access to residents living in Jamaica Bay watershed neighborhoods (Van Hooreweghe, 2012; also see chapter 6), and communities affected by Hurricane Sandy (Architecture for Humanity, New York, 2013).

Culturally and Ecologically Valued Aspects of the Jamaica Bay Watershed SES

In the Jamaica Bay watershed, there remains a lack of research that investigates culturally and ecologically valued factors from an integrated SES perspective. Current research has trended more toward human impacts on the estuarine ecosystem, and very few publications have considered the nested relationship of the social system as part of the larger ecological system of the JB watershed. Therefore, the work of the Jamaica Bay watershed researchers is to integrate social science research with other forms of research. Similarly, researchers in other ecosystems will have to take stock of the existing data to understand what kinds of data are missing.

Drivers that Affect the Jamaica Bay Watershed SES

Drivers that are of particular interest to social research that have not been examined include influences of the global economy and housing (low-income and public housing, in particular). Overall, the literature was found to include a significant, but incomplete, view of the drivers in the SES of the Jamaica Bay, indicating further research is needed to identify key drivers from the social-ecological perspective.

Key Indicators from an SES Perspective

There is a primary need to have baseline measures of the resilience of the SES of a watershed. We recognize that resilience measures must be inferred indirectly as surrogates because of their complexity (Carpenter et al., 2005). We suggest those working on resiliency issues in urban estuaries use a slightly adapted version of the Cutter et al. (2008) disaster resilience of place (DROP) model for measuring community resilience. The DROP model includes analysis of ecological (i.e., erosion rates and biodiversity), social (i.e., demographics, social networks, community values (including culturally valued aspects of the watershed), and community organizations that can advocate for equitable resource distribution (Adger et al., 2005), economic (i.e., employment), institutional (i.e., hazard mitigation plans), infrastructure (i.e., critical infrastructure, housing stock, and transportation), and community competence (i.e., community awareness of potential environmental change).

Concluding Thoughts

For those working in urban estuaries to pursue integrated SES resilience research, we suggest that future research questions and approaches about resilience and adaptation be co-constructed with residents and members of community organizations. In a participatory way, the research can contribute to an understanding of resilience, while also enhancing networks and communication channels for resilience by engaging key stakeholders in the process. In the case of the Jamaica Bay watershed, there is a dearth of information on the social aspects of the indicators described above (but see chapters 6 and 11) and therefore a need to prioritize which and how those indicators are addressed in inquiry.

If data gathered through a participatory process were coupled with the robust existing ecological data (such as marsh acreage loss, erosion rates, and so on), we would begin to have a framework that captures an SES perspective. For this reason, we suggest that urban estuarine resilience research bring social science in conversation with ecological or biological science and as participatory endeavors between community members and researchers. The Jamaica Bay estuarine ecosystem, and others like it around the globe, have been dramatically affected by the human systems that surround them. For this reason, an SES resilience perspective is well positioned for acceptance precisely because it establishes that social systems and natural systems do not exist separate from one another (Berkes and Folke, 1998).

References

Adger, W.N., Hughes, T.P., Folke, C., Carpenter, S.R., and Rockström, J. 2005. Social-ecological resilience to coastal disasters. *Science* 309(5737): 1036–1039.

Aldrich, D.P. 2012. *Building Resilience: Social Capital in Post-Disaster Recovery*. Washington,

D.C.: Island Press.

Anthony, A., Atwood, J., August, P., Byron, C., Cobb, S., et al. 2009. Coastal lagoons and climate change: Ecological and social ramifications in U.S. Atlantic and Gulf coast ecosystems. *Ecology and Society* 14(1): 8.

Architecture for Humanity, New York. 2013. The (Post-Sandy) Neighborhood Assessment Project. Accessed at: http://newyork.architectureforhumanity.org/.

Armitage, D., and Johnson, D. 2006. Can resilience be reconciled with globalization and the increasingly complex conditions of resource degradation in Asian coastal regions? *Ecology and Society* 11(1): 2. Accessed at: http://www.ecologyandsociety.org/vol11/iss1/art2/.

Benotti, M.J., Abbene, I., and Terracciano, S.A. 2007. Nutrient Loading in Jamaica Bay, Long Island, New York: Predevelopment to 2005. *U.S. Geological Survey Scientific Investigations Report,* 2007-5051.

Berkes, F., and Folke, C. 1998. Linking social and ecological systems for resilience and sustainability. In: *Linking Social and Ecological Systems,* Berkes, F., and Folke, C. (eds.). 1–25. Cambridge: Cambridge University Press.

Binder, C.R., Hinkel, J., Bots, P.W.G., and Pahl-Wostl, C. 2013. Comparison of frameworks for analyzing social-ecological systems. *Ecology and Society* 18(4): 26.

Black, F.R. 1981. Jamaica Bay: A history, Gateway National Recreation Area, New York-New Jersey, Historic Resources Study. Washington, D.C.: U.S. Department of Interior, National Park Service, North Atlantic Regional Office, Division of Cultural Resources; Cultural Resources Management Study No. 3.

Boyle, K. 2002. *Breaking the Waves: Rockaway Rises and Rises Again.* Scotts Valley, CA: Rising Star Press.

Breton, Y., Brown, D., Davy, B., Haughton, M., and Ovares, L. 2006. *Coastal Resource Management in the Wider Caribbean: Resilience, Adaptation, and Community Diversity.* Kingston, Jamaica: Ian Randle Publishers.

Buckley, P.A., and McCarthy, M.G. 1994. Insects, vegetation, and the control of laughing gulls *(Larus atricilla)* at Kennedy International Airport, New York City. *Journal of Applied Ecology* 31(2): 291–302.

Bures, R., and Kanapaux, W. 2011. Historical regimes and social indicators of resilience in an urban system: The case of Charleston, South Carolina. *Ecology and Society* 16(4): 16.

Burger, J. 1983. Jet aircraft noise and bird strikes: Why more birds are being hit. Environmental Pollution Series A, *Ecological and Biological* 30(2): 143–152.

Burger, J. 1985. Factors affecting bird strikes on aircraft at a coastal airport. *Biological Conservation* 33(1): 1–28.

Burger, J. 2000. Consumption advisories and compliance: The fishing public and the deamplification of risk. *Journal of Environmental Planning and Management* 43(4): 471–488.

Carpenter, S.R., Westley, F., and Turner, M.G. 2005. Surrogates for social ecological systems. *Ecosystems* 8: 941–944.

Cicin-Sain, B., Knecht, R.W., Jang, D., and Fisk, G.W. 1998. *Integrated Coastal and Ocean Management: Concepts and Practices.* Washington, D.C.: Island Press.

Colten, C.E., and Giancarlo, A. 2011. Losing resilience on the Gulf Coast: Hurricanes and social memory. *Environment: Science and Policy for Sustainable Development* 53(4): 6–19.

Conway, F.D., and Pomeroy, C. 2006. Evaluating the human—as well as the biological—objectives of cooperative fisheries research. *Fisheries* 31(9): 447–454.

Cutter, S.L., Barnes, L., Berry, M., Burton, C., Evans, E., Tate, E., and Webb, J. 2008. A place-based model for understanding community resilience to natural disasters. *Global Environmental Change* 18(4): 598–606.

Folke, C., Hahn, T., Olsson, P., and Norberg, J. 2005. Adaptive governance of social-ecological systems. *Annual Review of Environment and Resources* 30: 441–473.

Friesinger, S., and Bernatchez, P. 2010. Perceptions of Gulf of St. Lawrence coastal communities confronting environmental change: Hazards and adaptation, Québec, Canada. *Ocean and Coastal Management* 53(11): 669–678.

Gateway National Recreation Area, 2007. An Update on the Disappearing Salt Marshes of Jamaica Bay, New York. Brooklyn, NY: National Park Service, U.S. Department of the Interior, Gateway National Recreation Area.

Goldin, M. 1993. Effects of Human Disturbance and Off-Road Vehicles on Piping Plover Reproductive Success and Behavior at Breezy Point, Gateway National Recreation Area. Thesis: University of Massachusetts, Amherst.

Gornitz, V., Couch, S., and Hartig, E.K. 2002. Impacts of sea level rise in the New York City metropolitan area. *Global and Planetary Change* 32(1): 61–88.

Gotham, K.F., and Campanella, R. 2013. Constructions of resilience: Ethnoracial diversity, inequality, and post-Katrina recovery, the case of New Orleans. *Social Sciences* 2(4): 298–317.

Hartig, E.K., Gornitz, V., Kolker, A., Mushacke, F., and Fallon, D. 2002. Anthropogenic and climate-change impacts on salt marshes of Jamaica Bay, New York City. *Wetlands* 22(1): 71–89.

Hartig, E.K., Mushacke, F., Fallon, D., and Kolker, A. 2000. A Wetlands Climate Change Impact Assessment for the Metropolitan East Coast Region. Metropolitan East Coast (MEC) Regional Assessment, Center for International Earth Science Information Network, Earth Institute, Columbia University.

Hartley, T.W., and Robertson, R.A. 2008. Stakeholder collaboration in fisheries research: Integrating knowledge among fishing leaders and science partners in northern New England. *Society and Natural Resources* 22(1): 42–55.

Hildebrandt, E. 2005. Double Trauma in Belle Harbor: The Aftermath of September 11th and November 12th, 2001 in the Rockaways. In: *Wounded City, the Social Impact of 9/11*, Foner, N. (ed). New York: Russell Sage Foundation.

Jacob, K.H., Gornitz, V., and Rosenzweig, C. 2007. Vulnerability of the New York City metropolitan area to coastal hazards, including sea-level rise: Inferences for urban coastal risk management and adaptation policies. In: *Managing Coastal Vulnerability: Global, Regional, Local,* McFadden, L., Nicholls, R., and Penning-Rowsell, E. (eds). 139–156. New York: Elsevier Publishers.

Kaplan, I.M. 1998. Regulation and compliance in the New England conch fishery: A case for co-management. *Marine Policy* 22(4–5): 327–335. doi:10.1016/S0308-597X(98)00048-7.

Kaplan, L., and Kaplan, C.P. 2003. *Between Ocean and City, The Transformation of Rockaway.* New York: Columbia University Press.

Kaplan, I.M., and McCay, B.J. 2004. Cooperative research, co-management and the social dimension of fisheries science and management. *Marine Policy* 28(3): 257–258. doi:10.1016/j.marpol.2003.08.003

Kenward, A., Yawitz, D., and Raja, U. 2013. Sewage Overflows from Hurricane Sandy. Climate Central. Available at: http://www.climatecentral.org/pdfs/Sewage.pdf.

Kornblum, W. 1983. Racial and cultural groups on the beach. *Ethnic Groups* 5: 109–124.

Kornblum, W., and Van Hooreweghe, K. 2010. Jamaica Bay Ethnographic Overview and Assessment. Boston: Northeast Region Ethnography Program, National Park Service.

Kurlansky, M. 2007. *The Big Oyster: History on the Half Shell*. New York: Random House.

Larsen, R.K., Calgaro, E., and Thomalla, F. 2011. Governing resilience building in Thailand's tourism-dependent coastal communities: Conceptualising stakeholder agency in social-ecological systems. *Global Environmental Change* 21(2): 481–491.

Low, S.M. 2005. *Rethinking Urban Parks: Public Space and Cultural Diversity*. Austin, TX: University of Texas Press.

Millennium Ecosystem Assessment. 2005. *Ecosystems and Human Well-Being: Synthesis*. Washington, D.C.: Island Press.

Morrow, B.H. 2008. *Community Resilience: A Social Justice Perspective* (Vol. 4). Oak Ridge, TN: CARRI Research Report.

Nursey-Bray, M., and Harvey, N. 2013. Bridging the science-policy divide in the coastal zone: Is there a role for learning processes? In: *Global Challenges in Integrated Coastal Zone Management*, Moksness, E., Dahl, E., and Støttrup, J. (eds). 218–228. Chicester, West Sussex: John Wiley & Sons, Ltd.

Olsson, P., Folke, C., and Berkes, F. 2004. Adaptive co-management for building resilience in social-ecological systems. *Environmental Management* 34(1): 75–90.

Plummer, R. 2009. The adaptive co-management process: An initial synthesis of representative models and influential variables. *Ecology and Society* 14(2): 24.

Sorensen, J. 1997. National and international efforts at integrated coastal management: Definitions, achievements, and lessons. *Coastal Management* 25(1): 3–41.

Svendsen, E.S., Baine, G., Northridge, M.E., Campbell, L.K., and Metcalf, S.S. 2014. Recognizing resilience. *American Journal of Public Health* 104(4): 581–583.

Swanson, R.L., and Wilson, R.E. 2008. Increased tidal ranges coinciding with Jamaica Bay development contribute to marsh flooding. *Journal of Coastal Research* 24(6): 1565–1569. doi:10.2112/07-0907.1

Tett, P., Sandberg, A., Mette, A., Bailly, D., Estrada, M., Hopkins, T.S., d'Alcala, M.R., and McFadden, L. 2013. Perspectives of social and ecological systems. In: *Global Challenges in Integrated Coastal Zone Management*, Moksness, E., Dahl, D., and Støttrup, J. (eds.) 229–243. Chichester, West Sussex: John Wiley & Sons, Ltd.

Tidball, K.G., Krasny, M.E., Svendsen, E., Campbell, L., and Helphand, K. 2010. Stewardship, learning, and memory in disaster resilience. *Environmental Education Research* 16(5–6): 591–609.

Turner, N.J., Davidson-Hunt, I.J., and O'Flaherty, M. 2003. Living on the edge: Ecological and cultural edges as sources of diversity for social-ecological resilience. *Human Ecology* 31(3): 439–461.

Van Hooreweghe, K.L. 2012. The Creeks, Beaches, and Bay of the Jamaica Bay Estuary: The Importance of Place in Cultivating Relationships to Nature. Dissertation. New York, NY. The Sociology Department. The Graduate Center, CUNY.

Waldman, J. 2008. Research Opportunities in the Natural and Social Sciences at the Jamaica Bay Unit of Gateway National Recreation Area. National Park Service Jamaica Bay Institute.

Walker, R.M. 2006. Innovation type and diffusion: An empirical analysis of local government. *Public Administration* 84: 311–335.

Walker, B., and Salt, D. 2006. *Resilience Thinking: Sustaining Ecosystems and People in a*

Changing World. Washington, D.C.: Island Press.

Walker, B.D., and Salt, D. 2012. *Resilience Practice: Building Capacity to Absorb Disturbance and Maintain Function*. Washington, D.C.: Island Press.

Wigand, C., Roman, C.T., Davey, E., Stolt, M., Johnson, R., Hanson, A., Watson, E.B., Moran, S.B., Cahoon, D.R., Lynch, J.C., and Rafferty, P. 2014. Below the disappearing marshes of an urban estuary: Historic nitrogen trends and soil structure. *Ecological Applications* 24: 633–649. doi:10.1890/13-0594.1

Williamson, T. 2013. Beyond social capital: Social justice in recovery and resilience. *Risk, Hazards and Crisis in Public Policy* 4(1): 28–31.

Wilson, L., and Wiber, M.G. 2009. Community perspectives on integrated coastal management: Voices from the Annapolis Basin area, Nova Scotia, Canada. *Ocean and Coastal Management* 52(11): 559–567.

PART II

Social-Ecological Systems of Jamaica Bay

4

Dynamics of the Biophysical Systems of Jamaica Bay

Larry Swanson, Michael Dorsch, Mario Giampieri,
Philip Orton, Adam S. Parris, and Eric W. Sanderson

Jamaica Bay is often a wondrous place of serenity and even solitude surrounded by a bustling metropolis of millions of people. In its seemingly remote stillness, one can catch glimpses of the Manhattan skyline and watch massive airplanes appear to float into John F. Kennedy International Airport. And although is appears wild and beautiful, little of the bay's physical setting hasn't been altered or manipulated over the past 150 years, including its geomorphology, and the sources of its freshwaters, its sediments, and its marshes. It is our "National Urban Estuary—a Bay of Contrasts" (Swanson, 2007). Resilience requires adopting a nested view of how social-ecological systems interact, and the widest and broadest nest is that of the rocks, sediments, waters, and energies of Jamaica Bay.

Jamaica Bay has been shaped by a long continuum of geological, geomorphological, and oceanographic events, some of which trace a history hundreds of millions of years in the making and others of which continue to change minute by minute in the bay today. From a physical systems perspective, resilience is as much about accommodating ongoing change as it is trying to hold the system in place; in fact, as we shall explore, some of the reasons why Jamaica Bay has become less resilient over time include human efforts to stabilize the bay in a particular form or to use it for particular purposes. Urbanization, land fill, coastal engineering efforts, and pollution have all left their mark on Jamaica Bay and its watershed. In this chapter we briefly review salient features of the physical system, including geographical setting, geology, soils, water, climate, sea level rise, and land use and development. These conditions set the stage for the ecological and social interactions described in the next two chapters.

Geographical Setting

Jamaica Bay is the westernmost of the coastal lagoons that characterize the south shore of Long Island in New York State, in the northeastern part of the United States of America, lying at approximately 40.6178° N, 73.8425° W. The bay is physically separated from the Atlantic Ocean by the barrier island known as the Rockaway Peninsula. The mouth of the bay connects with the sea through Rockaway Inlet at its western end (color plate V).

Jamaica Bay, as with other estuaries, should be thought of as including not only the waters in the bay, but also the lands that drain into the bay, including its topographic watershed and its "sewershed" (figures 4-1, 4-2, and 4-3). The latter is the area from which waters are transported to the bay via sewer and stormwater pipes, resulting in combined

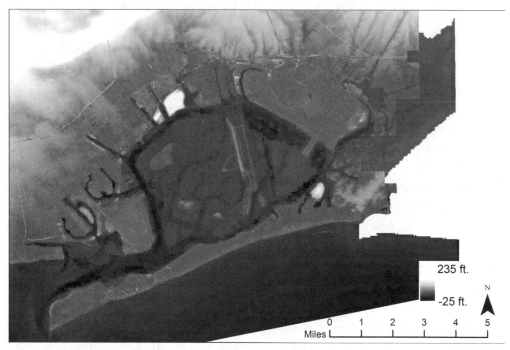

Figure 4-1. The current topography and bathymetry of the Jamaica Bay watershed. Although Jamaica Bay is a low-lying spot on the coastal plain, it does have a topography all its own. This digital elevation model (DEM) shows the Jamaica Bay watershed with the water drained away. Zero is set to the current mean sea level. The navigation channels and "borrow pits" for sediments used for extending the land are clearly visible as darker depths, and as are the sanitary landfills (garbage dumps), all now closed, are the bright white oblongs, by far the highest features on the bay's margins. This map is courtesy of Mario Giampieri of the Wildlife Conservation Society. The terrestrial data is based on the New York City 1-foot DEM (data.cityofnewyork.us/City -Government/1-foot-Digital-Elevation-Model-DEM-/dpc8-z3jc), and the bathymetric data is the Federal Emergency Management Agency DEM modified by Philip Orton of the Stevens Institute of Technology. The base map is from Esri, HERE, DeLorme, MapmyIndia, © OpenStreetMap contributors, and the GIS user community.

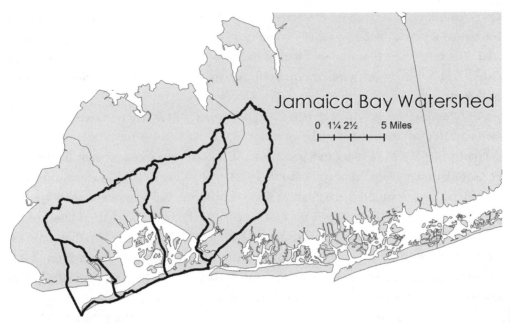

Figure 4-2. The topographic watersheds of Jamaica Bay. This map shows the four topographic watersheds draining into the bay on the western end of Long Island, New York, based on analysis of the modern topography. Courtesy of Joy Cytryn of Hunter College, City University of New York.

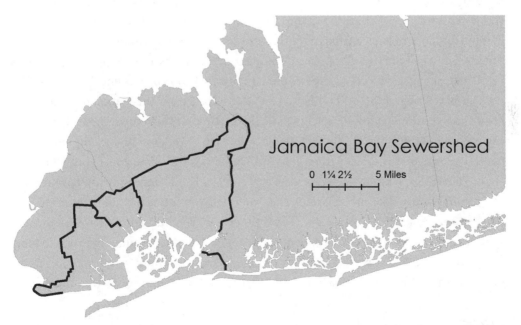

Figure 4-3. The sewersheds of Jamaica Bay. This map shows the areas that drain to the four wastewater treatment plants managed by the City of New York, which mainly, but not entirely, follow the topographic watersheds. Two other plants treat sewage from Nassau County (not shown). Courtesy of Joy Cytryn of Hunter College, City University of New York.

sewer overflows that may draw waters (and pollution) from beyond the topographic watershed (as described below). The combined Jamaica Bay watershed is highly urbanized. The catchment lies primarily within the New York City boroughs of Brooklyn and Queens, and includes the eastern portion in the Town of Hempstead in Nassau County, New York.

In total, the Jamaica Bay watershed is approximately 91,000 acres (36,900 ha). Open water and wetlands currently make up about 13,000 acres (5,300 ha) of this space (figure 4-1; color plate V). New York City's portion of the watershed is approximately 53,000 acres (21,450 ha) and includes residential, commercial, and industrial lands; vacant and underdeveloped lands; landfills; marshlands; and JFK Airport. The airport occupies approximately 4,300 acres (1,740 ha) on the eastern edge of the bay. About 1.6 million people are served by the four wastewater treatment plants (WWTPs) on the bay that drain the New York City portion of the sewershed (Interstate Environmental Commission, 2011). Interestingly, Kornblum and Van Hooreweghe (2010) reported the 2008 population of the watershed as about 2 million, up from about 1.9 million in 1970.

Eight highly altered and channelized tributaries (Sheepshead Bay, Paerdegat Basin, Fresh Creek, Hendrix Creek, Spring Creek, Shellbank Basin, Bergen Basin, and Thurston Basin), modifications of the historic streams that once flowed into Jamaica Bay, exist as stubby inlets of water into the surrounding urbanized matrix. The sewershed feeds combined sewage and storm flows into four WWTPs that have been constructed on the margins of Jamaica Bay at Jamaica, and in the 26th Ward, Rockaway, and Coney Island. These facilities in turn discharge treated effluent to tributaries and the open waters of the bay. The Coney Island WWTP releases effluent primarily on ebb current.

The bay is slightly flood-dominated, as determined from the National Oceanic and Atmospheric Administration's (NOAA's) tidal current tables and the U.S. Geological Survey water level station in Rockaway Inlet near the Marine Parkway Bridge, but more or less neutral elsewhere. Historically the mean depth of the bay was about 3 feet (1 m); today it is about 16 feet (5 m) (figure 4-1; West-Valle et al., 1991). This change in depth has increased the volume of the tidal prism (volume of water between mean low tide and mean high tide), so that today it is about 8.4×10^4 m^3 (Robert Wilson, School of Marine and Atmospheric Sciences, Stony Brook University, pers. comm.). Residence time (time elapsed for a particle to exit the region of its initial position as determined from hydrodynamic modeling) now varies from hours to about a month, depending on location within the bay.

Geology

Long Island's metamorphic bedrock formations of gneiss, schist, and marble are more than 400 million years old (Precambrian age) and are topped with overlying sands and clays deposited after erosion from the uplifted New England Upland in the west during the Cretaceous period about 70 million years ago (Buxton and Shernoff, 1995; Mills, 1974; USDA NRCS, 2001). These deposits range from near the surface in northwestern Queens to about 1,250 feet (350 m) deep at the Queens/Nassau County boundary beneath Rockaway Beach, suggesting a dip of the bedrock of approximately 74 feet per mile (14 m/km) under western Long Island (Misut and Voss, 2007). Bedrock porosity is estimated to be less than 1 percent (Misut and Voss, 2007).

The Cretaceous deposits of Long Island have been repeatedly buried by glaciations during the Pleistocene period (Sirkin, 1996). At some point, a pre-Illinoian–aged glacier from either the late Pliocene or early Pleistocene advanced near New York City and diverted the drainage of the proto-Hudson and Pensauken trunk rivers to the east, north of Staten Island and across Brooklyn (Buxton and Shernoff, 1995; Stanford, 2010; Stanford and Harper, 1991). The proto-Hudson river exited through a south-trending valley in Queens that likely opened an early version of what we now call Flushing Meadows and Jamaica Bay. This valley was subsequently filled by glacially deposited sediments.

The main and last glacial event shaping Jamaica Bay occurred during the Wisconsin stage of the Pleistocene, reaching its maximum extent approximately 21,000 years ago (Stanford, 2010). A massive continental glacier extended into the New York City region from lobes in the Hudson River Valley and Connecticut River Valley and crossed the region now known as New York City (Sirkin, 1996). The moraine that formed at the toe of the glacier is the Harbor Hill moraine, which created the rolling topography north and west of Jamaica Bay, including the neighborhoods of Bay Ridge, Park Slope, and Crown Heights, in Brooklyn, and Richmond Hill, Kew Gardens Hills, and Hillcrest, in Queens (New York City Department of Environmental Protection [NYCDEP], 2007).

Jamaica Bay falls entirely south of the terminal moraine on the outwash plain of this former glacier. The flat peneplain is made of sorted sediments, including sands, silts, and gravels washed out from the glacier and repeatedly worked by braided stream channels on the surface, which are still partially visible in the historic and modern topography (figure 4-1; color plate I). The glaciers began to retreat from the New York City region approximately 20,000 years ago (Stanford, 2010), leading to a long period of glacio-isostatic adjustment, as the land surface rebounded (and continues to rebound) upward after release from the glacial weight. For areas slightly south of the moraine, the result has been a slight decrease in ground elevation, as the underlying mantle has shifted under the New York/New Jersey coastal plain (Kemp and Horton, 2013; Engelhart and Horton, 2012). The combination of glacio-isostatic adjustment and the melting of glacial ice waters accounts for the particular history of sea level rise in the New York City region, as described below (McHugh et al., 2010).

The multiple glacial episodes over the last 1.8 million years left layers of stratified drift in the form of several distinct layers of gravel and clay that underlay western Long Island.

Jameco Gravel overlies the Magothy Formation and is the oldest deposit of the Pleistocene era. The brown gravel consists of igneous, metamorphic, and sedimentary material.
The relatively coarse sediments were probably deposited by meltwater from glaciation. It
is several hundred feet thick under the bay (Buxton and Shernoff, 1995). Gardiners Clay
lies on top of the Jameco Gravel. It was most likely deposited between glaciations. It is not
present around Barren Island and Far Rockaway (Buxton and Shernoff, 1995).

In the postglacial period, the area's geology has been redefined by water and wind erosion and depositions that have eroded sediments and soils from some areas and deposited
them in other areas (Mills, 1974). As sea level rose as the last glaciers melted, wave action
and littoral drift have created barrier islands and sand splits such as Coney Island and the
Rockaway Peninsula in the area, and offshore winds have piled sand up into dunes along
the coastline. The long-term drift of sand is from east to west, as can be seen from the
shapes of the barrier islands along Long Island's south shore (Yasso and Hartman, 1975).
In turn the Rockaway barrier island protects the interior of Jamaica Bay from direct wave
action. This loss of wave action, in addition to changes in sedimentology, is hypothesized
to have led to the development of interior wetland islands in the late eighteenth or early
nineteenth century (see color plates I–III; Sanderson, in review).

During the late nineteenth century, the Rockaway Peninsula was growing an average of 253 feet per year (77 m/yr) to the west based on analysis of historical U.S. Coast
Survey charts (i.e., Hassler et al., 1844; Bache et al., 1861; Bache et al., 1882; U.S. Coast
and Geodetic Survey, 1910; M. Giampieri and E. Sanderson, Wildlife Conservation Society, pers. comm.) (figure 4-4). This period of rapid growth and dynamism is attributed
to the influence of littoral drift moving from east to west along the south shore of Long
Island (Taney, 1961). The Jamaica Bay barrier beach system, the westernmost of a series
of barrier beach and lagoon systems spanning 99 miles (160 km) from Montauk to Coney
Island, receives sediment eroding from the Ronkonkoma terminal moraine at the eastern end of Long Island (Hess, 1987b). Furthermore, Hess (1987a) recorded a positive
linear relationship between average annual storm energies and annual net changes in
the sediment budget in the natural barrier system, despite sediment shifting between the
beach face and shoals. Sediment input from littoral drift, coupled with wave action and
extreme storm events, historically produced a dynamic system of shifting beaches and
barrier islands. This natural replenishment of sediment from longshore sediment drift
was interrupted by the construction of engineered structures in the twentieth century.
Groin construction began in 1922 along the Rockaways, and a jetty was built at Breezy
Point, on the western tip of the peninsula, in 1933 (Psuty et al., 2010). The U.S. Army
Corps of Engineers (USACE) built eight stone groins and a stone bulkhead at Fort Tilden
in 1943. As of this writing, there are at least twenty-eight groins along the coast from
Jacob Riis Park to Fort Tilden.

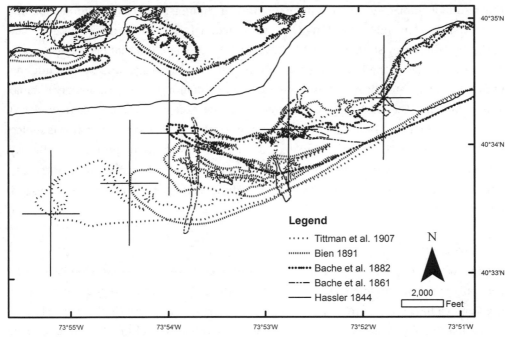

Figure 4-4. The Rockaway Peninsula advances, 1844–1907. Charts made by the U.S. Coast Survey have been georeferenced and overlaid to show the extension of the Rockaway spit barrier island in the late nineteenth century. On average over this 63-year period, a proto-version of Breezy Point was created. The tip of the peninsula (indicated by the crosses) grew on average 266 feet (81 meters) per year to the west and 79 feet (24 meters) per year to the south. In other words, coastal processes made approximately 2.5 miles (3.8 km) of new land during this time. Courtesy of Mario Giampieri of the Wildlife Conservation Society.

In addition to structures designed to stabilize the coast, during the twentieth century the USACE became an important geomorphological agent by supplying sand from dredging of the Rockaway Inlet to Jamaica Bay and from offshore areas to the beaches (Psuty et al., 2010) and more recently for marsh restoration efforts. Kana (1995) estimated that more than 25 million cubic yards (19 million m³) of material were placed on Rockaway beaches starting in the 1920s. USACE (1973) estimated that 3.8 million cubic yards (2.9 million m³) of material were placed along Coney Island at various locations throughout the 1920s as well.

Soils

Jamaica Bay contains extensive but now declining tidal marshes (see discussion in chapter 5) that have resulted in organic deposits over glacial outwash. Common soil orders include Ipswich, Pawcatuck, Matunuck, and Sandy Hook types (USDA NRCS, undated). Many of the soils found along the shoreline in the Jamaica Bay region today are anthropogenic, the result of human development in the area, including residential, commercial, and industrial development, landfilling with waste materials, and dredging operations (New York City Soil Survey Staff, 2005; NYCDEP, 2007). Waste materials that permeate the soils along the fringes of the bay include ash from coal, wood, garbage, and bone rendering. Other dumped materials found are construction and demolition debris, dredge material, and cellar dirt. The landscape in the area (with the exception of several, now closed, landfills; see figure 4-1) is naturally gently sloping down to sea level, so the area is subject to tidal flooding (USDA NRCS, undated). In eastern and northern portions of the bay, fine sand is the primary sediment type. In the southern and western sections of the bay, where high energy currents and waves occur, sediments range from fine to medium sands (USFWS, 1997).

Freshwater

Historically approximately a dozen freshwater streams flowed into Jamaica Bay. They were generally short-run streams that connected directly to tidal creeks in the salt marshes lining Jamaica Bay. Some of the most important historic streams were Hook Creek, Hassock (or Thurston's) Creek, and Gerritsen Creek. Some of the creeks had small ponds associated with them—notably the Beaver Pond where beavers are said to have lived—and also Hill Pond, Springfield Pond, and Conselyea's Pond. Many of the streams were dammed in the historic period for milling (Black, 1981). Although remnants of some creeks and ponds still exist in parks near the bay, most have been channelized in the lower portions or paved over and piped under roads and development in upper sections.

Four freshwater aquifers underlay the Jamaica Bay area (Lloyd Aquifer, Magothy Aquifer, Jameco Aquifer, and Upper Glacial Aquifer) (Misut and Voss, 2007). The freshwater in these aquifers comes from rainfall and snowmelt that infiltrate and percolate through porous areas such as lawns, undeveloped lands, parks, and cemeteries, and through seepage from the bottoms of lakes, ponds, and streams into the fine and coarse sands and silts. Urban development has resulted in increases in the number of impervious surfaces, such as buildings, roads, sidewalks, and other paved or constructed surfaces, that inhibit some of the infiltration and percolation from recharging the aquifers. Much of the water is diverted directly into Jamaica Bay through the storm sewer system (NYCDEP, 2007). The aquifer system in Jamaica Bay over the last 150 years has been under pressure from increased demand from the growing population in the area for residential, commercial, and industrial freshwater consumption and the decreasing ability of the aquifers to recharge due to impervious surfaces and sewer diversion. In the early twentieth century, Brooklyn and Queens both increased pumping from the aquifers in the region to the point that concerns were raised over saltwater intrusion (Buxton and Shernoff, 1995). By 1950, Brooklyn had ended much of its pumping of the aquifers and Queens significantly decreased its consumption from them beginning in the 1970s. Today the aquifer system underlying Jamaica Bay has begun to rise with a resulting increase in the elevation of the groundwater table, but the quality of the groundwater has deteriorated due to saltwater intrusion, road salt, leakage from sewer lines, and spills from chemical and petroleum products (NYCDEP, 2007). We are constantly reminded of the former importance of these water supply resources around the bay by names such as Aqueduct Racetrack and Conduit Avenue.

Marine Waters

Jamaica Bay itself can be described as a temperate, seasonally eutrophic estuary with water salinities in open waters ranging from approximately 20 to 26. Water temperatures vary seasonally from 32.6°F (1°C) to 46.4°F (26°C). The acidity ranges from 6.8 to 9 on the pH scale (USFWS, 1997). NYCDEP monitors water chemistry at various locations throughout the bay. Of concern are nitrogen, dissolved oxygen, chlorophyll a, and pathogens, along with silica and organic carbon, which are consequences of wastewater treatment facilities and combined sewer overflows (NYCDEP, 2007; also see updates NYCDEP, 2010, 2012, 2014).

Many of the freshwater and brackish creeks that drain into Jamaica Bay have been bulkheaded and channelized. Two-thirds of the freshwater runoff is now diverted through the four sewage treatment facilities that produced approximately 287 million gallons per day (1.086×10^9 l/day) of secondarily treated effluent in the 1990s (Waldman, 2008; West-Valle et al., 1991). By 2010, the discharge rate had decreased to 238 million gallons per day (9.00×10^6 l/day) (Interstate Environmental Commission, 2011), largely from New York City water conservation measures. Currently the discharge rate is 223 million gallons per day (8.44×10^8 l/day) (John McLaughlin, New York City Department of Environmental Protection, pers. comm.). With the decreasing discharge, the nitrogen load from sewage treatment plants has dropped about 29 percent from 1990 levels, to about 36,000 pounds per day (16,330 kg/day) in the early 2000s, and is currently at 26,100 pounds per day (11,839 kg/day) (McLaughlin, pers. comm.). The New York State Department of Environmental Conservation goal is for the wastewater dischargers to Jamaica Bay to further reduce nitrogen loads by 20,000 pounds per day (9,072 kg/day) by 2020. Current daily freshwater inputs into the bay are approximately 0.5 percent of total volume of the bay (Rhoads et al., 2001).

High concentrations of nitrogen are a major factor contributing to the low levels of dissolved oxygen available in Jamaica Bay (NYCDEP, 2007). Algae are simple plants that proliferate and grow off of an abundance of the macronutrients nitrogen and phosphorus. Nitrogen is the limiting nutrient in Jamaica Bay, and because of anthropogenic increases in its concentration, algal growth is excessively stimulated. As the algae die and are decomposed by bacteria, the oxygen in the water is consumed, leading to hypoxia, or low concentrations of dissolved oxygen. Although there is year-to-year variability in dissolved oxygen concentrations, many of the NYCDEP monitoring stations indicate that summertime bottom oxygen concentrations have been increasing in the bay in the decade between 1995 and 2005 as a consequence of nitrogen reduction measures implemented by the city (NYCDEP, 2007).

Chlorophyll a concentrations are also an important water quality indicator in Jamaica Bay. Because algae produce chlorophyll a to capture the light needed for photosynthesis, it can be used as an indicator of the amount of algae present in the water. Chlorophyll a levels, which have been increasing since the early 1990s (NYCDEP, 2007), suggest that the water is eutrophic. Fecal coliform and enterococci concentrations, indicators of pathogenic contamination, are also monitored, and recently both have been at or below the levels set by the U.S. Environmental Protection Agency as acceptable for bathing (NYCDEP, 2007). They are currently not low enough to safely allow harvest of shellfish.

Tide and Storm Surge Regime

The average tide range in 2007 at Norton Point (southeastern Jamaica Bay) was 5.4 feet (1.65 m), relative to that of Sandy Hook, NJ at 4.7 feet (1.43 m). Tides are dominantly of the semidiurnal character, meaning that there are two highs per day. However, a small diurnal modulation averaging 0.5 feet (0.15 m) also exists, so one high each day is slightly larger than the other, and one low slightly lower than the other.

Superimposed on these tides are storm surges, most commonly cool-season "extra-tropical cyclones," but also occasional hurricanes and tropical storms (table 4-1). Extra-tropical cyclones occur at varying intensity and have historically caused storm tides (tide plus surge) of up to 9.7 feet (2.96 m) above mean lower low water (MLLW), which occurred in December 1992 (at New York Harbor's Battery; Orton et al., submitted). Hurricanes are less common here than nor'easters, but can provide threat of much higher storm tides, as exemplified by Hurricane Sandy's 14.1-foot (4.30 m) peak at the Battery and 13.9-foot (4.24 m) peak at Norton Point. Storm tides during Sandy at Rockaway Inlet near Floyd Bennett Field and at the Inwood gauging station (adjacent to JFK Airport) were 13.71 feet (4.17 m) and 13.90 feet (4.24 m) above MLLW, respectively. Looking at the flood hazard from any type of storm, and taking the lowest and highest estimates from recent studies (see summary in Orton et al., submitted), the 10-year return period storm tide at the Battery is 8.6 feet (2.62 m) (Zervas, 2013) to 9.1 feet (2.77 m) MLLW (Lopeman et al., 2015), and the 100-year storm tide is 10.6 feet (3.23 m) (Zervas, 2013) to 14.1 feet (4.30 m) (FEMA, 2014). The storm tide hazard for Jamaica Bay is similar to that of New York Harbor and Sandy Hook, with some events being larger in Jamaica Bay, but most are smaller (e.g., Sandy). The bay's interior is protected from ocean swells. During these storms, waves are created locally by the wind and are typically only a few feet in height, but nevertheless capable of causing moderate erosion.

Table 4-1. A chronology of severe storms, 1635–2012.

Date	Description
25 August 1635[1]	Hurricane
8 September 1667[1]	A "severe storm"
25 August 1693[2]	Accomack storm of October 1693
14 October 1706[2]	"great ruins here and mighty floods"
24 October 1716[2]	"gale"
19 October 1749[2]	"violent gale"
9 September 1769[2]	"violent gale"
13 October 1778[2]	"a violent gale at NNE with heavy rain"
9 October 1783[2]	"uncommon high tide attended with a hard gale at northwest"
24 September 1785[2]	"a heavy equinoctial storm"
19 August 1788[2]	"a severe gale . . . with incredible fury"
9 October 1804[2]	"a violent circulation"
24 August 1806[2]	The Great Coastal Hurricane
23 September 1815[2]	"very heavy rain and gales"
3 September 1821[2]	Norfolk and Long Island Hurricane
11–13 September 1878[3,4]	Storm type unknown; moderate severity
23–25 October 1878[3,4]	Hurricane; moderate severity
9–11 December 1878[3,4]	Hurricane; moderate severity
3 February 1880[3,4]	Low-energy extratropical storm
24 November 1885[3,4]	Storm type unknown; moderate severity
8–9 January 1886[3,4]	Storm type unknown; moderate severity
11–14 March 1888[3,4]	The "Blizzard of '88"; moderate severity nor'easter
6–12 September 1888[3,4]	Hurricane; low severity
12 August 1890[3,4]	Extratropical storm
1 March 1892[3,4]	Extratropical storm
21 April 1893[3,4]	Extratropical storm
15–26 August 1893[3,4]	"West Indian Cyclone" Hurricane
27 December 1894[3,4]	Extratropical storm
26 January 1895[3,4]	Extratropical storm
6 February 1895[3,4]	Extratropical storm
24 September 1897[3,4]	Other tropical storm
24–25 October 1897[3,4]	Extratropical storm
19 October 1898[3,4]	Extratropical storm
8 February 1899[3,4]	Extratropical storm
11–12 September 1900[3,4]	Extratropical storm
10–15 October 1900[3,4]	Other tropical storm
4 December 1900[3,4]	Extratropical storm
24 November 1901[3,4]	Extratropical storm
16–17 September 1903[3,4]	Hurricane
8–11 October 1903[3,4]	Extratropical storm/hurricane
14–15 September 1904[3,4]	Hurricane
25 January 1905[3,4]	Extratropical storm
4 March 1909[4]	Extratropical storm
12 February 1910[4]	Extratropical storm
30–31 August 1911[4]	Hurricane
15–16 September 1912[4]	Hurricane

Table 4-1. continued

26 December 1913[4]	Extratropical storm
1 March 1914[4]	Extratropical storm
7 December 1914[4]	Extratropical storm
4 April 1915[4]	Tropical storm
4–5 August 1915[4]	Tropical storm
20–21 July 1916[4]	Hurricane
10 August 1917[4]	Tropical storm
24 October 1917[4]	Extratropical storm
15 January 1918[4]	Extratropical storm
11–12 April 1918[4]	Extratropical storm
5 February 1920[4]	Extratropical storm
30 September 1920[4]	Hurricane
28–29 January 1922[4]	Extratropical storm
22–24 October 1923[4]	Hurricane
25–26 August 1924[4]	Hurricane
3–4 December 1925[4]	Hurricane
10 February 1926[4]	Extratropical storm
20 February 1927[4]	Extratropical storm
24 August 1927[4]	Hurricane
4 October 1927[4]	Tropical storm
12–13 August 1928[4]	Hurricane
19 September 1928[4]	Hurricane
16 April 1929[4]	Extratropical storm
2–3 October 1929[4]	Hurricane
30 January 1930[4]	Extratropical storm
3–5 March 1931[4]	Extratropical storm
13 May 1932[4]	Tropical storm
16 September 1932[4]	Tropical storm
9–10 November 1932[4]	Extratropical storm
27 January 1933[4]	Extratropical storm
23–24 August 1933[4]	Hurricane
17–18 September 1933[4]	Hurricane
19 June 1934[4]	Hurricane
8–9 September 1934[4]	Hurricane
17 November 1935[4]	Extratropical storm
7 February 1936[4]	Extratropical storm
18–19 September 1936[4]	Hurricane
1 October 1936[4]	Tropical storm
17 October 1936[4]	Extratropical storm
21 September 1938[4]	Long Island Express Hurricane
22–24 October 1938[4]	Tropical storm
28–29 October 1938[4]	Extratropical storm
19 August 1939[4]	Tropical storm
25–26 November 1939[4]	Extratropical storm
24 January 1940[4]	Extratropical storm
20 February 1940[4]	Extratropical storm
18 August 1940[4]	Hurricane
1–2 September 1940[4]	Hurricane

Table 4-1. continued

3 March 1942[4]	Extratropical storm
30 September 1943[4]	Tropical storm
26 October 1943[4]	Extratropical storm
4–5 January 1944[4]	Extratropical storm
3 August 1944[4]	Hurricane
14 September 1944[4]	Hurricane
21 October 1944[4]	Hurricane
30 November 1944[4]	Extratropical storm
16 January 1945[4]	Extratropical storm
26–27 June 1945[4]	Hurricane
18–19 September 1945[4]	Hurricane
22–29 November 1945[4]	Extratropical storm
6 December 1945[4]	Extratropical storm
29 May 1946[4]	Extratropical storm
9–10 November 1947[4]	Extratropical storm
26 December 1947[4]	Extratropical storm
5–6 October 1948[4]	Extratropical storm
24 August 1949[4]	Hurricane
20–21 August 1950[4]	Hurricane
11–12 September 1950[4]	Hurricane
25 November 1950[4]	Extratropical storm
8 December 1950[4]	Extratropical storm
30–31 March 1951[4]	Extratropical storm
25–26 April 1951[4]	Extratropical storm
4 February 1952[4]	Tropical storm
11–12 May 1952[4]	Extratropical storm
1–2 September 1952[4]	Hurricane
20–22 November 1952[4]	Extratropical storm
14–15 August 1953[4]	Hurricane
7 September 1953[4]	Hurricane
21 September 1953[4]	Tropical storm
23 October 1953[4]	Extratropical storm
6–7 November 1953[4]	Extratropical storm
3 May 1954[4]	Extratropical storm
31 August 1954[4]	Hurricane Carol
11 September 1954[4]	Hurricane Edna
15 October 1954[4]	Hurricane Hazel
13 August 1955[4]	Hurricane Connie
18–19 August 1955[4]	Hurricane Diane
20 September 1955[4]	Hurricane
14–16 October 1955[4]	Extratropical storm
9–11 January 1956[4]	Extratropical storm
16 March 1956[4]	Extratropical storm
6–7 October 1957[4]	Extratropical storm
27–28 February 1958[4]	Extratropical storm
19–22 March 1958[4]	Extratropical storm
29 August 1958[4]	Hurricane Daisy
28 September 1958[4]	Hurricane Helene

Table 4-1. continued

7 December 1959[4]	Extratropical storm
28–29 December 1959[4]	Extratropical storm
13–14 February 1960[4]	Extratropical storm
18–19 February 1960[4]	Extratropical storm
29–30 July 1960[4]	Tropical storm Brenda
12 September 1960[4]	Hurricane Donna
12 December 1960[4]	Extratropical storm
13–14 January 1961[4]	Extratropical storm
16 January 1961[4]	Extratropical storm
3–4 February 1961[4]	Extratropical storm
13 April 1961[4]	Extratropical storm
16 April 1961[4]	Extratropical storm
21 September 1961[4]	Hurricane Esther
21–23 October 1961[4]	Extratropical storm
6–8 March 1962[4]	Extratropical storm
28–29 August 1962[4]	Hurricane Alma
27–28 September 1962[4]	Extratropical storm
10 November 1962[3]	Extratropical storm
6–7 December 1962[3]	Extratropical storm
29 October 1963[3]	Hurricane
6–7 November 1963[3]	Extratropical storm
29–30 November 1963[3]	Extratropical storm
13 January 1964[3]	Extratropical storm
14 September 1964[3]	Hurricane
17 January 1965[3]	Extratropical storm
22–24 January 1966[3]	Extratropical storm
31 January 1966[3]	Extratropical storm
21 September 1966[3]	Extratropical storm
28–29 December 1966[3]	Extratropical storm
7 February 1967[3]	Extratropical storm
29 April 1967[3]	Extratropical storm
24–26 May 1967[3]	Extratropical storm
16–17 September 1967[3]	Hurricane
4 December 1967[3]	Extratropical storm
28–29 May 1968[3]	Extratropical storm
10–13 June 1968[3]	Hurricane
12–13 November 1968[3]	Extratropical storm
9–10 September 1969[3]	Hurricane
17 November 1969[3]	Hurricane
10–11 December 1969[3]	Extratropical storm
25–28 December 1969[3]	Extratropical storm
10 February 1970[3]	Extratropical storm
29 March 1970[3]	Extratropical storm
2–3 April 1970[3]	Extratropical storm
10–13 November 1970[3]	Extratropical storm
19 November 1970[3]	Extratropical storm
16–17 December 1970[3]	Extratropical storm

Table 4-1. continued

7–9 February 1971[3]	Extratropical storm
3–4 March 1971[3]	Extratropical storm
6–7 April 1971[3]	Extratropical storm
27–28 August 1971[3]	Tropical storm
11–14 September 1971[3]	Tropical storm
25 January 1972[3]	Extratropical storm
3–5 February 1972[3]	Extratropical storm
13 February 1972[3]	Extratropical storm
19–20 February 1972[3]	Extratropical storm
20–25 June 1972[3]	Tropical storm
8–9 November 1972[3]	Extratropical storm
14–15 November 1972[3]	Extratropical storm
15–16 December 1972[3]	Extratropical storm
26–29 January 1973[3]	Extratropical storm
26 October 1973[3]	Tropical storm
29 October 1973[3]	Extratropical storm
9 December 1973[3]	Extratropical storm
16–17 December 1973[3]	Extratropical storm
21 December 1973[3]	Extratropical storm
22–23 February 1974[3]	Extratropical storm
10 March 1974[3]	Extratropical storm
21 March 1974[3]	Extratropical storm
30 March 1974[3]	Extratropical storm
1–2 December 1974[3]	Extratropical storm
3–5 April 1975[3]	Extratropical storm
22–27 September 1975[3]	Extratropical storm
10 August 1976[5]	Tropical storm Belle
24 September 1985[5]	Tropical storm Henri
27 September 1985[5]	Tropical storm Gloria
29–30 September 1988[5]	Tropical storm Chris
13 July 1996[5]	Tropical storm Bertha
16–17 September 1999[5]	Tropical storm Floyd
19–20 September 2000[5]	Extratropical storm Gordon
7 September 2008[5]	Tropical storm Hanna
28 August 2011[5]	Tropical storm Irene
28–29 October 2012[5]	Hurricane Sandy

[1] Roth and Cobb, 2001.

[2] Ludlum, 1963.

[3] Hess and Harris, 1987a.

[4] USACE, 1973.

[5] NOAA Hurricane Research Division, 2015.

Source: Courtesy of Mario Giampieri and Christopher Spagnoli of the Wildlife Conservation Society.

Table 4-1. This table documents a severe storm chronology, no doubt incomplete, for the Jamaica Bay watershed, compiled from the sources listed.

The morphology of Jamaica Bay has been heavily modified by human activities, particularly over the last 150 years, extensively changing the elevations above and below the water line (figure 4-5). The average low tide depth in the bay was approximately 3 feet (0.91 m) prior to modern development, which includes landfilling of shallows, channel dredging, bulkheading, and the removal of sediments from "borrow" pits. The average low tide depth has increased to approximately 16 feet (4.88 m) today. Dredging activities alone are estimated to account for nearly 70 percent of the increased volume in the bay (Rhoads et al., 2001). The dredging has also caused tidal amplification between the ocean entrance and the head of the bay. In the vicinity of Rockaway Channel that amplification in tidal range is about 1 foot (0.30 m) and at Norton Point 1.6 feet (0.49 m) (Swanson and Wilson, 2008). Combining the effects of regional relative sea level rise and the increases in tidal ranges equates to mean high water being higher today relative to a century ago by about 1.3 feet (0.40 m) at Barren Island and 1.5 feet (0.46 m) at Norton Point (Swanson and Wilson, 2008). Amplification of the tidal range has both increased

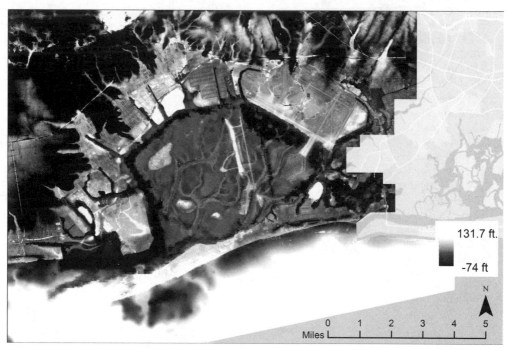

Figure 4-5. Effects of landfill and dredging, c. 1609–2014. This map is a subtraction of a reconstruction of the pre-European topography from the modern topography (the DEM shown in figure 4-1). Positive values (light gray and white colors) indicate areas where landfill has occurred; negative values (dark gray and black) indicate areas of subsidence, dredging, and land removal. Courtesy of Mario Giampieri of the Wildlife Conservation Society. The historical topography and bathymetry is from the Welikia Project. The modern data are from the sources given in the caption of figure 4-1.

the amount of flooding and extended flooding duration throughout the tidal cycle over the bay's marshes. Morphological changes have also increased the average residence time of a water molecule in the northern and eastern portions of the bay, extending this time from approximately 11 to 33 days (NYCDEP, 1994).

Climate

The climatological setting of Jamaica Bay is defined by a humid continental climate moderated by its proximity to the Atlantic Ocean (NYCDEP, 2007). During the winter months, dry and cold arctic air masses extend from the northern interior of the United States and Canada into the region. Meanwhile, warm, humid air masses are carried from the Gulf of Mexico on winds out of the south and southwest in the summer months. During warmer months of the year, a third air mass carries cooler, humid air into the region, moderating temperatures and bringing cool, cloudy, and damp weather conditions (NYCDEP, 2007). The average annual air temperature for Jamaica Bay from 1949 to 2000 was 53.9°F (12.2°C), with an average annual high temperature for the time period of 61.0°F (16.1°C) and an average annual low temperature of 46.7°F (8.2°C). Annual average temperatures are getting hotter in the region, with a rate of increase of approximately 0.3°F (0.17°C) per decade over the measurement period. Between 1900 and 2013, average air temperatures in New York City increased 3.4°F (1.9°C). Global climate models predict annual average temperatures to continue to increase by 4.1–4.7°F (2.3–2.6°C) through the 2050s and by 5.3–8.8°F (2.9–4.9°C) by the 2080s over the current temperature baseline (NPCC, 2015; see chapter 8). Extreme temperature events in the form of heat waves are projected to triple in the region by the 2080s, and extreme cold events are expected to decrease (NPCC, 2015).

Average annual precipitation between 1949 and 2000 was 40.16 inches (102 cm) (NOAA NCEI, 2015). There are no distinct wet or dry seasons in Jamaica Bay, but less precipitation tends to occur in the winter due to the drier arctic air masses that dominate. Monthly winter snowfall accumulations can reach 3–10 inches (7.6–25.4 cm) on average in the region, but coastal nor'easters can occasionally bring recurring snowfalls that exceed 20 inches (50.8 cm) of accumulation (NYCDEP, 2007). Average annual precipitation has increased alongside increases in average annual temperatures with a rate of increase of approximately 0.8 inch (2.0 cm) per decade and a total of 8 inches (20.3 cm) between 1900 and 2013. Global climate models suggest that annual average precipitation is expected to continue to increase in the region by 4–11 percent by the 2050s and 5–13 percent by the 2080s (NPCC, 2015). In addition to coastal nor'easters, the region is susceptible to tropical storms and hurricanes, of which Hurricanes Sandy and Irene serve as the most recent examples. Under global climate models, extreme precipitation days are projected to increase with one and a half times more events possible per year by the 2080s as compared with the current climate (NPCC, 2015).

Sea Level Rise

Relative sea level rise is another climatological factor facing the Jamaica Bay region now and into the future. Sea level in the New York City region has averaged an increase of 1.2 inch (3.0 cm) per decade, or 1.1 feet (0.3 m) since the year 1900 (figure 4-6). This rate is nearly double the observed global rate of 0.5–0.7 inches (1.3–1.8 cm) per decade over a similar time period. About 40 percent of the city's rate is due to subsidence from isostatic readjustment (NPCC, 2015). In the future, sea levels are projected under global climate models to continue to rise by about 11–21 inches (27.9–53.3 cm) by the 2050s, by an additional 18–39 inches (45.7–99.1 cm) by the 2080s, and potentially by 6 feet (1.8 m) by 2100 (NPCC, 2015). Based on estimated sea level rise alone, increased frequency and intensity of coastal flooding in Jamaica Bay is virtually a certainty. Global climate models project an approximate doubling of the frequency and intensity of coastal flooding and a ten- to fifteen-fold increase in the frequency of the current 100-year coastal flood by the 2080s (NPCC, 2015). Compounded with possible increases in the frequency and intensity of tropical cyclones, the flooding potential could be even higher than the sea level projections alone account for (NPCC, 2015).

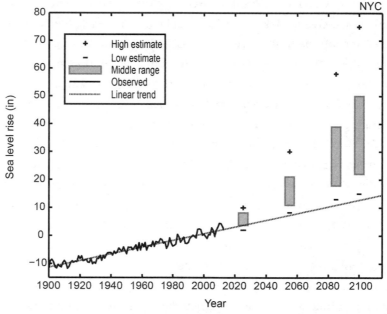

Figure 4-6. Sea level rise at the Battery, 1900–2014, with predictions for the future. Mean sea level in New York Harbor has risen at an average rate of 0.11 inches/year (2.83 mm/year) between 1856 and 2004, significantly faster than the average global sea level rise of 0.07 inches/year (1.7 mm/year) between 1870 and 2004. Future predictions are based on downscaled global climate change models as described in chapter 8. Adapted from Horton et al. (2015). Courtesy of Dan Bader at NASA Goddard Institute for Space Studies.

Land Use and Development

Prior to the arrival of western European colonists, indigenous inhabitants of Jamaica Bay were part of the Lenape people of the Lenapehoking territory that extended from the Lower Hudson Valley, across western Long Island where Jamaica Bay is located, and south along the Delaware River watershed (Carman, 2013). The Lenape people who lived in Jamaica Bay would have likely spoken the Munsee dialect and formed two closely related groups, the Canarsie of the western shores of Jamaica Bay and the Rockaway of the eastern shores (Black, 1981). The Canarsie and Rockaway peoples lived in small permanent settlements along the tidal streams and bay and engaged primarily in subsistance fishing and hunting, and especially shellfishing, using dug-out canoes and wooden rakes (Black, 1981; color plate I).

In the mid-1600s, Dutch colonists were the first to meet with Lenape chiefs on Long Island and soon followed the first sale of lands by the Canarsie to the Dutch. By the late 1600s, very few survivors of the Canarsie or Rockaway peoples remained in the Jamaica Bay area (Black, 1981). The towns of Jamaica, Flatlands, New Lots, and Flatbush were founded as early Dutch villages in the area (color plate II; Black, 1981). From the time of the Dutch colonization of the area until the mid-1800s, nearly all those who lived in the communities surrounding Jamaica Bay pursued farming and agriculture for subsistence and light sale to markets in New York (color plate III). However, exploitation of Jamaica Bay's abundance of fin- and shellfish provided supplemental livelihoods for locals and later brought nonlocal fishermen to the area in search of clams, mussels, and fishes, such that the town of Jamaica passed regulations in the late 1700s that required fishermen from outside of Jamaica to pay a shilling for every thousand shellfish taken from the bay (Black, 1981).

Beginning in the mid-1800s, the area began to undergo industrial transformations with manufacturing establishments specialized in creating fertilizers and fish oils sited on Barren Island (color plate III). Mill Island became the site of a large lead smelting plant that processed 4,000 tons (3,629 metric tonnes) of ore and produced 3,800 tons (3,447 mt) of solder, tin, and lead annually, which was transported via Jamaica Bay's water channels. During this time period, the Jamaica Bay area also began to be a site for processing and disposing of refuse from New York City (Black, 1981). More residents moved to the area to work in factories, and ships transported manufactured goods and catches from the fisheries back to New York markets (Black, 1981). However, navigation was a challenge. The 1904 coast pilot for Jamaica Bay stated that the bay was "so full of marsh islets and islands as to render navigation utterly impossible except to very light-draft vessels" (U.S. Coast and Geodetic Survey, 1904). Today most of those marshes have been lost to development and other factors (see chapter 5).

By the turn of the twentieth century, disposal of New York's dead animals and garbage became one of the primary industrial activities in Jamaica Bay, with 500–1,000 tons (454–907 metric tonnes) of refuse arriving daily to Barren Island, such that a 1912 newspaper article refered to the operations as "among the largest of their kind in the world" (Black, 1981; see also Miller, 2000). The small cove behind Barren Island, now buried under Floyd Bennett Field, became known as Dead Horse Bay. Bulkheading to fill island marshes on Mill Island was put into place in the early 1900s, creating 332 acres (134 ha) of new uplands, and in 1915 a 100-foot-wide (30.5 m) channel was dredged from Jamaica Bay channel to Mill Basin to transport goods manufactured at Mill Basin's factories (Black, 1981). Bergen Island, which is now Bergen Beach, also experienced bulkheading and dredging operations, that transformed the landscape and created additional upland acreage. Dredging and filling operations in the 1920s and 1930s also reformed the area such that Barren Island merged with Riches Meadows to become part of what is now Floyd Bennett Field, and the areas of Sand Bay around Canarsie were bulkheaded and filled in and the Canarsie Pier was built (Black, 1981). Much more ambitious plans were made for Jamaica Bay's "improvement," as suggested by color plate IV. Although those plans were never fulfilled, the legacy of waste disposal and shoreline hardening has transformed the outwash plain, making the bay an undulating ribbon of asphalt and landfill, with some hillocks rising 30–40 feet (9–12 m) above sea level, clearly visible on figures 4-1 and 4-5. The landfills at Pennsylvania Avenue, Fountain Avenue, and Edgemere pose a potential for restored grassy natural resource areas, but are also sources of pollution leaching into Jamaica Bay (Benotti et al., 2007).

The beaches of the Rockaway Peninsula became a popular recreation spot for New Yorkers in the nineteenth century, especially after the Civil War, when rail lines extended from Brooklyn to Canarsie Landing, where travelers could take ferries from the landing to Rockaway Peninsula. Hotels, restaurants, and saloons began to establish themselves in the area (Black, 1981), including the famous Marine Pavilion, an elite hotel in Far Rockaway constructed in 1833 and burnt down in 1864. Other venues soon followed, and in 1880, the New York, Woodhaven, and Rockaway Railroad began operations to Rockaway Peninsula using a 4.8-mile (7.7 km) trestle across Jamaica Bay (Black, 1981). These railways would be replaced by paved streets and highways and the subway in the twentieth century.

After the Civil War, the commercial fishing industry in Jamaica Bay also experienced significant growth, with many of the fishermen living and operating out of Canarsie (see discussion of the oyster fishery in chapter 7). By the turn of the twentieth century, the waters of Jamaica Bay were becoming so polluted that the fish and shellfish were contaminated and often caused serious illnesses when consumed; the pollution problem resulted in the shellfishing industry ceasing operations in the bay in 1926 (Black, 1981). By the 1930s, many of the recreational developments in Canarsie and other locations around the bay began to decline as industrial activity increased and the residential populations around the bay began to grow (Black, 1981; Swanson et al., 1992). In the 1950s, the subway system took over and rebuilt the railroad trestle that crossed the bay to Rockaway Peninsula, and with it came a residential housing boom around the bay (Black, 1981). The area of Broad Channel became one of the most densely built-up sites in the early 1900s, first with seasonal fishing stations and dwellings, but then with a rapidly growing permanent residential community (Black, 1981). By 1930, competing visions were in place for how the area of Jamaica Bay should be developed. It was in 1930 that urban planner and later park commissioner Robert Moses announced a plan to build a parkway from Brooklyn to connect with the Cross Island Parkway to turn much of Jamaica Bay into a large city park. Construction of the Belt Parkway was completed in 1936 with four lanes that were widened to six after the 1930s (Black, 1981). The Marine Parkway Bridge was also completed in 1937 and connected the western tip of Rockaway Peninsula with Flatbush Avenue and the Belt Parkway (Black, 1981).

Floyd Bennett Field was opened as New York City's first municipal airport in 1931 and was used through the 1940s by the U.S. Navy as one of the busiest naval air stations in the United States during World War II (NPS, 2015). Up until the early 1970s, the navy continued to use Floyd Bennett Field as part of the Naval Air Reserve system, but it was decommissioned in 1971 and soon became part of the Gateway National Recreation Area (NPS, 2015). In the early 1940s, New York City was looking to expand its airport options beyond that of LaGuardia Airport, and a recreation site in the north of Jamaica Bay occupied by Idlewild Golf Course was selected. Originally planned to be a 1,000-acre airport, Idlewild Airport (today JFK Airport) was five times that size when it opened in July 1948 and began commercial flights (The Port Authority of New York and New Jersey, 2015a). Today, JFK airport is one of the biggest international hubs in the United States with more than eighty airlines operating from its six airline terminals and 9 miles (14.5 km) of total runway length (The Port Authority of New York and New Jersey, 2015b). All of these different historical factors have led to considerable land fill on the margins of the bay, changing its physical shape and character (figure 4-1).

Today, the Jamaica Bay watershed represents one of the most heavily urbanized regions in the United States with high if variable population densities (color plate V). Along the coastline on Rockaway Peninsula and in the residential areas of Broad Channel, New Howard Beach, Old Howard Beach, and Hamilton Beach, more than 130,000 people inhabit the neighborhoods closest to the water in 24,200 buildings, of which 96 percent are residential and contain 53,000 housing units (New York City Special Initiative for Rebuilding and Resiliency, 2013). More than half of all housing units are located in multifamily buildings, which include six public housing developments operated by the New York City Housing Authority. Many of these are located in Far Rockaway and Rockaway, where the largest populations reside and at the highest densities. Far Rockaway, with

54,000 residents and 39 people per acre, has a population density close to the average for New York City overall (New York City Special Initiative for Rebuilding and Resiliency, 2013). Taking into account the broader watershed of Jamaica Bay, well over a million people live in the residential areas of Queens and Brooklyn that are located within the watershed (U.S. Census Bureau, 2014a, b).

For the future, resilience of Jamaica Bay requires a full understanding and appreciation of the dynamics that shape its physical systems. Over the long run it is futile to deny the ways that wind, water, and wave shape the bay. These trends are all the more serious because of ongoing climate change, which is raising sea levels and changing the basic climatology of not only Jamaica Bay, but also New York City and the world. The best hope for long-term resilience lies with finding ways for the ecological and human systems to live in concert with, and not in contradiction to, Jamaica Bay's lands and waters.

References

Bache, A.D., et al. 1861. Coast Chart No. 21. New York Bay and Harbor, New York. Washington, D.C.: Survey of the Coast of the United States. Accessed at: http://historical charts.noaa.gov/historicals/preview/.

Bache, A.D., Ferguson, J., Blunt, E., Whiting, H.L., Gilbert, S.A., et al. 1882. Coast Chart No. 20. New York Bay and Harbor, New York. Washington, D.C.: Survey of the Coast of the United States. Accessed at: http://historicalcharts.noaa.gov/historicals/preview/.

Benotti, M.J., Abbene, M., and Terracciano, S.A. 2007. Nitrogen Loading in Jamaica Bay, Long Island, New York, Predevelopment to 2005 (Scientific Investigations Report 2007-5051). U.S. Geological Survey, Reston, VA.

Bien, J.R., and Vermeule, C.C. 1891. Atlas of the Metropolitan District and adjacent country. Julius Bien & Co., New York. Accessed at: http://www.davidrumsey.com/.

Black, F.R. 1981. Jamaica Bay: A History. Cultural Resource Management Study 3. National Park Service, Washington, D.C. Accessed at: http://www.nps.gov/history/history/online_books/gate/jamaica_bay_hrs.pdf.

Buxton, H.T., and Shernoff, P.K. 1995. Ground-Water Resources of Kings and Queens Counties, Long Island, New York. U.S. Geological Survey Open-File Report 92-76. Accessed at: http://ny.water.usgs.gov/pubs/of/ofr9276/ofr9276.pdf.

Carman, A.E. 2013. *Footprints in Time: A History and Ethnology of the Lenape-Delaware Indian Culture*. Bloomington, IN: Trafford Publishing.

Engelhart, S.E., and Horton, B.P. 2012. Holocene sea level database for the Atlantic coast of the United States. *Quaternary Science Reviews* 54: 12–25. doi:10.1016/j.quascirev.2011.09.013.

FEMA. 2014. Region II Coastal Storm Surge Study: Overview, Federal Emergency Management Agency, Washington, D.C. Accessed at: https://data.femadata.com/NationalDisasters/Hurricane%20Sandy/RiskMAP/Public/Public_Documents/Storm_Surge_Reports/.

Hassler, F.R. 1844. Map of New-York Bay and Harbor and the Environs (Survey of the Coast of the United States). Washington, D.C.: U.S. Coast Survey. Accessed at: http://www.davidrumsey.com/.

Hess, L., and Harris, W.H. 1987a. Effect of storm energy and shoreline engineering on the sediment budget of a barrier beach, Rockaway, New York. *Northeastern Geology* 9: 110–115.

Hess, L., and Harris, W.H. 1987b. Morphology and sediment budget during recent evolution of a barrier beach, Rockaway, New York. *Northeastern Geology* 9: 94–109.

Horton, R., Little, C., Gornitz, V., Bader, D., and Oppenheimer, M. 2015. New York City Panel on Climate Change 2015 Report, Chapter 2: Sea Level Rise and Coastal Storms. *Annals of New York Academy of Sciences* 1336: 36–44. doi:10.1111/nyas.12593.

Interstate Environmental Commission. 2011. 2010 Annual Report. Accessed at: http://www.iec-nynjct.org/reports/2010.IEC.AnnualReport.pdf.

Kana, T.W. 1995. A mesoscale sediment budget for Long Island, New York. *Marine Geology* 126: 87–110. doi:10.1016/0025-3227(95)00067-9.

Kemp, A.C., and Horton, B.P. 2013. Contribution of relative sea-level rise to historical hurricane flooding in New York City. *Journal of Quaternary Science* 28: 537–541.

Kornblum, W., and Van Hooreweghe, K. 2010. Jamaica Bay Ethnographic Overview and Assessment. Northeast Region Ethnography Program, National Park Service, Boston, MA. Accessed at: http://macaulay.cuny.edu/eportfolios/adams2013/files/2013/08/Complete-NPS-report.pdf.

Lopeman, M., Deodatis, G., and Franco, G. 2015. Extreme storm surge hazard estimation in lower Manhattan. *Natural Hazards* 78(1): 1–37.

Ludlum, D.M., 1963. Early American Hurricanes 1492-1870. American Meteorological Society, Boston, MA.

McHugh, C.M., Hartin, C.A., Mountain, G.S., and Gould, H.M. 2010. The role of glacio-eustasy in sequence formation: Mid-Atlantic Continental Margin, USA. *Marine Geology* 277: 31–47. doi:10.1016/j.margeo.2010.08.009.

Miller, B. 2000. *Fat of the Land: Garbage of New York: The Last Two Hundred Years*. New York: Four Walls Eight Windows.

Mills, H.C. 1974. Geology of Long Island. Garvies Point Museum and Preserve. Accessed at: http://www.garviespointmuseum.com/geology.php.

Misut, P.E., and Voss, C.I. 2007. Freshwater–saltwater transition zone movement during aquifer storage and recovery cycles in Brooklyn and Queens, New York City, USA. *Journal of Hydrology* 337: 87–103. doi:10.1016/j.jhydrol.2007.01.035.

National Park Service (NPS). 2015. Detailed History of Floyd Bennett Field. Accessed at: http://www.nps.gov/gate/learn/historyculture/fbf.htm.

New York City Soil Survey Staff, 2005. New York City Reconnaissance Soil Survey. Staten Island, NY: United States Department of Agriculture, Natural Resources Conservation Service. Accessed at: http://www.soilandwater.nyc/uploads/7/7/6/5/7765286/reconnaissance_soil_survey_report.pdf.

New York City Special Initiative for Rebuilding and Resiliency. 2013. PlaNYC: A Stronger, More Resilient New York. New York: The City of New York. Accessed at: http://www.nyc.gov/html/sirr/html/report/report.shtml.

NOAA Hurricane Research Division, 2015. Continental United States Tropical Storms, 1851–1955, 1883–2014 (Revised to include 1951–1955). Atlantic Oceanographic & Meteorological Laboratory, Miami, FL.

NOAA NCEI (National Centers for Environmental Information). 2015. Climate at a Glance Time Series. 1895–2015. New York (JFK). Accessed at: http://www.ncdc.noaa

.gov/cag/time-series/us.

NPCC (New York City Panel on Climate Change). 2015. Building the knowledge base for climate resiliency. *Annals of the New York Academy of Sciences* (1336): 1–150.

NYCDEP (New York City Department of Environmental Protection). 1994. Jamaica Bay Comprehensive Watershed Management Plan. New York, NY.

NYCDEP. 2007. Jamaica Bay Watershed Protection Plan. Accessed at: http://www.nyc.gov/html/dep/html/harborwater/jamaica_bay.shtml.

NYCDEP. 2010. Jamaica Bay Watershed Protection Plan 2010 Update. Accessed at: http://www.nyc.gov/html/dep/pdf/jamaica_bay/jbwpp_update_10012010.pdf.

NYCDEP. 2012. Jamaica Bay Watershed Protection Plan 2012 Update. Accessed at: http://www.nyc.gov/html/dep/pdf/jamaica_bay/JBWPP_Update_100112_FINAL.pdf.

NYCDEP. 2014. Jamaica Bay Watershed Protection Plan 2014 Update. Accessed at: http://www.nyc.gov/html/dep/pdf/jamaica_bay/jbwpp_update_10012014.pdf.

Orton, P.M., Hall, T.M., Talke, S., Blumberg, A.F., Georgas, N., and Vinogradov, S. (submitted, 1/25/2016). A validated tropical-extratropical flood hazard assessment for New York Harbor, *Journal of Geophysical Research.*

The Port Authority of New York and New Jersey. 2015a. History of JFK International Airport. Accessed at: http://www.panynj.gov/airports/jfk-history.html.

The Port Authority of New York and New Jersey. 2015b. Facts and Information. Accessed at: http://www.panynj.gov/airports/jfk-facts-info.html.

Psuty, N.P., Dennehy, P., Silveira, T., and Apostolou, N. 2010. Coastal Geomorphology of the Ocean Shoreline, Gateway National Recreation Area: Natural Evolution and Cultural Modifications, a Synthesis (Natural Resource Report NPS/NERO/NRR - 2010/184). Fort Collins, CO: National Park Service.

Rhoads, J.M., Yozzo, D.J., Cianciola, M.M., and Will, R.J. 2001. Norton Basin/Little Bay Restoration Project. Historical and environmental background report. Report to U.S. Army Corps of Engineers.

Roth, D. and Cobb, H., 2001. Seventeenth Century Virginia Hurricanes. Accessed at: http://www.wpc.ncep.noaa.gov/research/roth/va17hur.htm.

Sanderson, E.W., 2016. Cartographic Evidence for Historical Geomorphological Change and Wetland Formation in Jamaica Bay, New York. *Northeastern Naturalist* 23: 277–304.

Sirkin, L. 1996. *Western Long Island Geology with Field Trips.* Watch Hill, RI: The Book & Tackle Shop.

Stanford, S.D. 2010. Onshore record of Hudson River drainage to the continental shelf from the late Miocene through the late Wisconsinan deglaciation, USA: Synthesis and revision. *Boreas* 39: 1–17. doi:10.1111/j.1502-3885.2009.00106.x.

Stanford, S.D., and Harper, D.P. 1991. Glacial lakes of the Lower Passaic, Hackensack, and Lower Hudson Valleys, New Jersey and New York. *Northeastern Geology* 13: 271–286.

Swanson, R.L. 2007. Jamaica Bay: The National Urban Estuary—Bay of Contrasts. Lecture prepared for a class, Waste Management Issues (MAR 392), School of Marine and Atmospheric Sciences, Stony Brook University.

Swanson, R.L., West-Valle, A.S., and Decker, C.J. 1992. Recreation vs. waste disposal: The use and management of Jamaica Bay. *Long Island Historical Journal* 5(1): 21–41.

Swanson, R.L., and Wilson, R.E. 2008. Increased tidal ranges coinciding with Jamaica Bay development contribute to marsh flooding. *Journal of Coastal Research* 24(6): 1565–1569.

Taney, N.E. 1961. Geomorphology of the South Shore of Long Island, New York (Technical Memorandum 128). Washington, D.C.: U.S. Army Corps of Engineers, Beach Erosion Board.

U.S. Army Corps of Engineers–New York District. 1973. Atlantic Coast of New York City from Rockaway Inlet to Norton Point, New York (Coney Island Area) Cooperative Beach Erosion Control and Interim Hurricane Study. New York: U.S. Army Corps of Engineers.

US Census Bureau. 2014a. Kings County, New York QuickFacts. Accessed at: http://quick facts.census.gov/qfd/states/36/36047.html.

US Census Bureau. 2014b. Queens County, New York QuickFacts. Accessed at: http:// quickfacts.census.gov/qfd/states/36/36081.html.

U.S. Coast and Geodetic Survey. 1904. United States Coast Pilot, part IV, 4th ed. Washington, D.C.: U.S. Coast and Geodetic Survey. 1910. Coast Chart No. 120. New York Bay and Harbor, New York. Washington, D.C.: U.S. Coast and Geodetic Survey. U.S. Government Printing Office. Accessed at: https://commons.wikimedia.org/wiki/File:1910_U.S._Coast_Survey_Nautical_Chart_or_Map_of_New_York_City_and_Har bor_-_Geographicus_-_NewYorkCity-uscs-1910.jpg.

USDA, NRCS (U.S. Department of Agriculture, Natural Resources Conservation Service). undated. Soil Survey of Gateway National Recreation Area, New York and New Jersey. Accessed at: http://www.soilandwater.nyc/uploads/7/7/6/5/7765286/gateway_soil_sur vey_report.pdf.

USFWS (U.S. Fish and Wildlife Service). 1997. Significant habitats and habitat complexes of the New York Bight watershed. U.S. Fish and Wildlife Service, Southern New England. New York Bight Coastal Ecosystems Program, Charlestown, RI.

Waldman, J. 2008. Research opportunities in the natural and social sciences at the Jamaica Bay Unit of Gateway National Recreation Area. National Park Service. Accessed at: http://www.nps.gov/gate/naturescience/upload/jbay-research%20opportunities.pdf.

West-Valle, A.S., Decker, C.J., and Swanson, R.L. 1991. Use impairments of Jamaica Bay. Waste Management Institute, Marine Sciences Research Center, State University of New York, Stony Brook, NY.

Yasso, W.E., and Hartman, E.M. 1975. Beach Forms and Coastal Processes. Albany, NY: New York Sea Grant Institute.

Zervas, C. 2013. Extreme Water Levels of the United States 1893–2010, NOAA Technical Report NOS CO-OPS 067, NOAA National Ocean Service Center for Operational Oceanographic Products and Services, Silver Spring, MD.

5

Ecology of Jamaica Bay: History, Status, and Resilience

Steven N. Handel, John Marra, Christina M. K. Kaunzinger,
V. Monica Bricelj, Joanna Burger, Russell L. Burke,
Merry Camhi, Christina P. Colón, Olaf P. Jensen,
Jake LaBelle, Howard C. Rosenbaum, Eric W. Sanderson,
Matthew D. Schlesinger, John R. Waldman,
and Chester B. Zarnoch

Coastal estuaries are renowned for their ecological diversity and abundance (e.g., Beck et al., 2003; Bertness, 2006). The same qualities that attract a variety of species draw people to settle on their shores and even build cities. Over time, these biologically rich environments can either be overwhelmed by human activities or relieved if people take proactive steps to conserve and restore estuaries and their watersheds, making them more resilient to environmental and human-induced change. These are challenges we address in this chapter reviewing the history, current status of, and prospects for resilience of Jamaica Bay's ecology.

The Jamaica Bay watershed in New York City is a model system for understanding the social-ecological interplay that stands at the heart of this book (Waldman, 2008; New York City Department of Environmental Protection, 2007; Black, 1981; chapters 1–3). The bay provides important habitats for resident and migratory animals, including species of high public interest, such as the diamondback terrapin, horseshoe crab, striped bass, and many shorebird and duck species (U.S. Fish and Wildlife Service, 1997). The bay also provides critical ecosystem goods and services for people (TEEB, 2011; see chapter 2, table 2-2), including reduction of the urban heat island effect, shoreline protection from storms, carbon storage, and nurseries and habitat for commercial and recreational fisheries. However, Jamaica Bay's ability to supply these ecosystem goods and services has been compromised by past and current actions and is challenged by future threats. Sadly, Jamaica Bay is not alone—we see it as representative of the kinds of resilience difficulties that urban estuaries throughout the region, the country, and the world face.

Management and restoration of degraded estuaries require an understanding of their long-term history. We describe patterns in the chronological drivers of ecological change in the bay and its watershed. Then we summarize the current status of representative biotic groups in the bay, organized from aquatic to upland habitats, emphasizing major stressors influencing ecosystem health. We close by summarizing knowledge gaps and research needs to address the resilience of Jamaica Bay's ecological systems.

Ecological History of the Jamaica Bay Watershed

Jamaica Bay has a long and complicated ecological history that stretches from the Pleistocene to the modern day. Here we briefly chart this trajectory, highlighting critical junction points, as a way of understanding the long-term drivers of change.

Drivers of Change, Pleistocene–1609

Prior to European colonization, Jamaica Bay was a natural landscape of extensive, interwoven habitats, including forests, wetlands, beaches, streams, and shallow marine waters. As climate warmed during the last 10,000 years, Long Island moved through a series of terrestrial ecological types, including spruce forest, pine forest, and eventually, around 5,000–6,000 years ago, to the mixed oak-hickory-chestnut forests (Sirkin, 1967) common on uplands at the time of European discovery. Sea level was also rising, moving landward from a shoreline approximately 100 miles (160 km) offshore at the height of the last glacial maximum (McHugh et al., 2010) to approximately its modern location, 2,000–4,000 years ago (Engelhart and Horton, 2012). Human habitation in the New York City region dates back approximately 8,000 years (Cantwell and Wall, 2001).

Drivers of Change, 1610–1910

The first European documentation of western Long Island comes from the voyage of Giovanni Verrazzano in 1524 (Wroth, 1970), who found a populated land surrounding a lake (typically thought to be Upper New York Harbor) surrounded by a favorable and beautiful landscape. Archeological evidence of the native Lenape people suggests patterns of shifting cultivation, hunting, and fishing throughout the region, and it is reasonable to expect similar activities around Jamaica Bay (Cantwell and Wall, 2001; Sanderson, 2009; color plate I).

The Dutch settled near the bay in the 1630s (Black, 1981) and were primarily interested in agriculture and cutting of salt hay in the tidal marshes for animal feed, which led to forest clearance in the watershed and some human disturbance within the bay itself (Peteet et al., 2008). Dutch and Native American populations also probably fished, hunted, and gathered shellfish, though as populations were low throughout the seventeenth and eighteenth centuries, these effects on the bay and its watershed were likely relatively minor (Black, 1981). Seventeenth-century maps show extensive fringing marshlands along the mainland edge of the bay, but do not document interior marshes (see chapter 4), probably because of the relative position and state of the sandy barrier island system, which appears to have been dynamic throughout this period.

Late eighteenth- and early nineteenth-century maps show the first appearance of interior marsh islands in Jamaica Bay, and by the mid-nineteenth century, they appear in approximately modern form (Sanderson, 2016). Detailed bathymetric surveys from 1877 to 1878 provide evidence of submerged vegetation (likely *Zostera maritima*) growing in the bay. Around the bay, the watershed was mostly agricultural land, with small woodlots until the early twentieth century.

The nineteenth century also saw a new use for Jamaica Bay, as a dump. Barren Island, an approximately 60-acre (24.3 ha) island near the Rockaway Inlet, was the site of several factories for rendering animal carcasses into ash and glue. Barren Island also received garbage from New York City, and eventually became, via landfill, Floyd Bennett Field (Miller, 2000).

Drivers of Change, 1911–1945

Rapidly growing population in the late nineteenth century drove agriculture out, giving way to suburban dwellings. An early twentieth-century plan to turn Jamaica Bay into the second major harbor of New York led to dredging of channels for shipping (Black, 1981). Hard edges to the bay were constructed, and fringing salt marshes at the bay's edge were filled in with dredge sands, construction debris, and household garbage to support the needs of maritime commerce and inland development (see Hendrick, 2006, for changes to the landform). The resulting channels caused an increase in the tidal range within the bay by up to 40 percent and an increase of the water volume of the bay by 350 percent (Swanson and Wilson, 2008). Examples of the building that occurred contemporaneously with the channel dredging include Canarsie Pier (1925), Cross Bay Boulevard (1923), and Floyd Bennett Field (1931).

Drivers of Change 1946–Present

Development expanded into the outlying parts of Brooklyn and Queens, converting more agricultural lands to houses, businesses, and transportation infrastructure. A new airport (now John F. Kennedy International Airport) was constructed on top of Idlewild Golf Course, which had been built on landfill. To make air travel safe, the Port Authority of New York and New Jersey manages bird populations on and near the airport (e.g., laughing gulls, see Brown et al., 2001a).

Development greatly increased how much of the land was covered by impermeable surface, increasing stormwater runoff into the bay, raising soil and air temperatures, and sharply curtailing sediment and natural freshwater inputs. The plant community of New York City also changed with hundreds of new species being introduced from abroad, and many populations of native species lost in the process of ecosystem conversion and fragmentation (Robinson et al., 1994; Clemants and Moore, 2005; Handel et al., 2013).

With urbanization also came new infrastructure for water treatment (New York City Department of Environmental Protection, 2009). By the early twenty-first century, New York City Department of Environmental Protection (NYCDEP) installed four wastewater treatment plants on the edge of Jamaica Bay, which processed 240 million gallons of wastewater per day, representing 99 percent of the freshwater inputs into the bay. Those inputs formerly included a daily discharge of approximately 34,800–39,700 pounds of nitrogen per day (15,800–18,000 kg/day), in contrast to estimated predevelopment inputs of approximately 78 pounds of nitrogen per day (35.6 kilograms/day) coming from streams (Benotti et al., 2007). High levels of nitrogen result in lowered dissolved oxygen in waterways, excessive algae growth, and decreased penetration of light. Recognizing these ecological impacts in the Jamaica Bay Watershed Protection Plan, New York City made commitments in 2010 to further decrease nitrogen input (NYCDEP, 2014). Currently, discharge rates are approximately 26,100 pounds of nitrogen per day (11,800 kg/day) averaged over a year.

Also in the twentieth century came recognition of Jamaica Bay's value for wildlife and recreation. In the 1950s, under Robert Moses, the New York City Parks Department took jurisdiction of Jamaica Bay and created a wildlife refuge (Black, 1981). Moses also took the opportunity to continue work on the "Circumferential Parkway," now known as the Belt Parkway. In 1972 most of Jamaica Bay came under control of the federal government to form the Jamaica Bay Wildlife Refuge as a unit of the Gateway National Recreation Area.

Climate change is expected to drive ecological alteration in Jamaica Bay. In New York City, there was a trend toward higher temperatures over the twentieth century, similar to a regional trend in the Northeast, suggesting that this is not just a result of the urban heat island effect (Horton et al., 2015a). Year-to-year precipitation has become more variable, especially since the 1970s (Horton et al., 2015a). Sea level rose an average of 1.2 inches (3 cm) per decade since 1900 (Horton et al., 2015b). Projections are for future warming over the 1971–2000 baseline, with a midrange average temperature increase of 5–10°F (3-6°C) by 2100, and increased and wider variation in precipitation (Horton et al., 2015a). Midrange sea level projections are for an increase of 22–50 inches (56–127 cm), with the high estimate of up to 75 inches (191 cm), by 2100 (Horton et al., 2015b), as discussed in chapter 4.

Current Characteristics of Ecological Communities of Jamaica Bay

The massive changes over the last 400 years (table 5-1; color plates I–V) have undermined the resilience of the ecological communities of Jamaica Bay, but a remarkable abundance of species are still found. Here we summarize the current status of a number of critical species, species groups, and communities, working from the waters of Jamaica Bay into upland areas.

Table 5-1. Land cover change in the Jamaica Bay watershed, c. 1609, c. 1877, and 2014.

Land/water cover class	Percentage of Jamaica Bay watershed		
	c. 1609	c. 1877	2011
Open water	62	56	50
Developed, high intensity	0	0	21
Developed, medium intensity	0	0	15
Developed, low intensity	0	0	4
Developed, open space	<1	<1	1
Cultivated crops	<1	16	<1
Barren land (rock/sand/clay)	1	2	1
Deciduous and mixed forest	26	8	1
Shrub/scrub	<1	<1	1
Grassland herbaceous	<1	0	1
Emergent herbaceous wetlands	11	17	4
Woody wetlands	<1	1	0
Miscellaneous	0	0	1
Total	100	100	100

Source: The Welikia Project, Wildlife Conservation Society

Table 5-1. Note that values indicate the percentage of the total area occupied by a given land cover type, not acreage. The 1609 land cover types come from the Welikia Project and are preliminary estimates; the 1877 data from analysis of U.S. Coast Survey charts; and the 2014 data from the same sources as color plate VI. Courtesy of Mario Giampieri of the Wildlife Conservation Society.

Phytoplankton

At the base of the estuarine food chain are phytoplankton, microscopic, free-floating algae that depend on light and are nourished by dissolved nutrients. The seasonal cycle of phytoplankton here is fairly typical of regional coastal estuarine systems, with a spring bloom largely composed of diatoms, which gives way to flagellated forms in the summer (Wallace and Gobler, 2015). The spring bloom can bring chlorophyll a levels of 60–100 mg m^{-3}, which define the bay as hypereutrophic. Changes in nutrient ratios (nitrogen:phosphorus:silicon) influence species composition and size structure of phytoplankton communities (Wallace and Gobler, 2015), leading to changes in the food supply for benthic (bottom) and pelagic (open water) grazers.

Submerged Aquatic Vegetation

The increased input of nitrogen to the bay's waters throughout most of the twentieth century (Benotti et al., 2007), plus the demise of oyster beds, has increased the water's turbidity, decreasing the light reaching the bottom. Nineteenth-century U.S. Coast Survey charts indicate eelgrass (*Zostera marina*), now gone, along the northern edge of Jamaica Bay and in Grassy Bay, but very little good data on historic patterns of eelgrass exist. Coupled with the deepening of parts of the bay, the current lack of light penetration severely limits submerged aquatic vegetation in favor of phytoplankton. Wallace and Gobler (2015) suggested that lack of light might limit growth of macroalga (*Ulva* sp.) ("sea lettuce") on the benthic surface.

Mollusks

Ribbed mussels (*Geukensia demissa*), an integral component of salt marshes, are currently the major suspension-feeding bivalves in Jamaica Bay, and particularly threatened by sea level rise along the edges of marshes. Ribbed mussels are highly tolerant of extreme environmental conditions (high temperature, sulfide in sediments), and in other estuaries have been shown to play important ecological roles in controlling phytoplankton biomass, influencing water quality via removal of nitrogen, reduction of turbidity (e.g., Jordan and Valiela, 1982) and bacteria (potentially human pathogens), and carbon cycling via use of detrital plant material as food (Kreeger and Newell, 2001). They are also an ideal tool for habitat improvement, because, unlike other commercially or recreationally exploited bivalves such as hard clams and oysters, they are not desirable as food and consequently pose no risk to human health after growing in contaminated waters.

Other bivalves such as oysters (*Crassostrea virginica*) and hard clams (*Mercenaria mercenaria*), which are not directly associated with salt marshes, also need to be considered in restoration efforts. In the late nineteenth century, Jamaica Bay supported abundant populations of these two bivalves, and at its peak produced up to 700,000 bushels (24, 670 m^3) of oysters per year (Franz, 1982). However, oysters are now either absent or extremely limited in abundance and distribution (Waldman, 2008). (See chronology and discussion in chapter 7.)

Studies examining the potential for oyster restoration in Jamaica Bay show that juvenile oysters transplanted to Jamaica Bay exhibit high growth rates and that adult oysters will successfully spawn (Zarnoch and Schreibman, 2012; Levinton et al., 2013). However, mature oysters experience abnormally high mortality, which challenges restoration efforts (Levinton et al., 2013; Hoellein and Zarnoch, 2014). The mortality appears to be associated with poor water quality conditions, but additional research is needed. Hard clams remain abundant, with densities up to $5/ft^2$ ($54/m^2$) observed in 1981–1982 (Franz and Harris, 1988). High densities of hard clams may be attributed to an absence of harvest pressure and to abundant food resources, while the lack of oysters is likely due to poor recruitment (no available substrate) or possibly eutrophication causing bacterial biofilms that dissuade larval settlement, as in Chesapeake Bay (Jackson et al., 2001). Diseases (e.g., MSX and Dermo) and oyster drills (*Urosalpinx cinerea*) play a significant role in high mortality rates (John McLaughlin, pers. comm., NYCDEP).

Horseshoe Crabs and Other Macroinvertebrates

The American horseshoe crab (*Limulus polyphemus*) relies on the remaining undisturbed sandy beaches to provide critical spawning habitat where they come ashore each spring by the thousands to lay, fertilize, and bury millions of eggs in the sand (figure 5-1). These eggs contribute to the carbon budget of this ecosystem (Botton and Loveland, 2011) and provide

Figure 5-1. An American horseshoe crab (*Limulus polyphemus*). These ancient creatures come ashore to lay their eggs on sandy beaches without coastal erosion protections such as bulkheads or sandbags. Courtesy of Robert Pos of the U.S. Fish and Wildlife Service.

a food supply for shore birds, some of which must quickly build up fat reserves to survive their own long-distance migrations (Gillings et al., 2007). Horseshoe crab hatchlings take up residence on the sandy tidal flats or silty tidal creeks, where they forage and grow. As the summer progresses, many juvenile crabs fall prey to predators (Botton et al., 2003).

It is not clear whether the small number of horseshoe crab hatchlings that survive to adulthood compensates for the thousands of adults that perish in the spring mating frenzy, but recruitment and spawning numbers overall both appear lower in urban sites compared with less urban locations (Mattei et al., 2010). Spawning habitat and nursery beaches are vital for survival, and both are experiencing rapid decline. Beach erosion and storm surge have required costly beach restorations at regular intervals. Although data indicate that the nourished beaches eventually support spawning adults and newly hatched juveniles (Botton et al., 2014), contravening bulkheading, sandbagging, and other shoreline hardening lead to habitat loss (Jackson et al., 2010).

The educational and ecotourism potential of beach ecosystems that support horseshoe crabs further increases the crabs' economic value (Walls et al., 2002). Through citizen science and civic engagement, local researchers and students are monitoring populations (Colón and Rowden 2014), trying to understand the steady decline in spawning adults (Faurby et al., 2010; Leschen and Correia, 2010).

The invertebrate community changes with water quality and habitat change, but also from the introduction of nonnative crustaceans and other taxa. Periwinkles (*Littorina littorea*) were introduced to the Northeast in the nineteenth century from Europe, and invasive crab species (*Carcinus maenas* and *Hemigrapsus sanguineus*) are having significant effects on marine communities (Grosholz and Ruiz, 1996; Kraemer et al., 2007). Additional species will likely arrive from ongoing commercial marine activities (Briski et al., 2012).

Fish

Surprisingly little is known about fish populations in Jamaica Bay (Waldman, 2008), given the importance of recreational fishing and the ecological role of fish as key predators and prey of other organisms (figure 5-2). Periodic fish surveys have occurred within the bay and its tributaries, but differences in methods generally preclude analysis of trends in fish abundance. The most comprehensive information on fish in Jamaica Bay is the ancillary data collected in the annual seine surveys of young striped bass from the Hudson River that occur in the western bays of Long Island, which includes considerable effort in Jamaica Bay (Socrates, 2010). More than one hundred species of fishes are known for the bay (Trust for Public Land and NYC Audubon, 1987. Nearby Barnegat Bay, New Jersey, also maintains a rich diversity of fish species (Szedlmayer

Figure 5-2. People (*Homo sapiens*) can be a useful a species too. Brooklyn College students seine for fish, hoping to discover some of the approximately one hundred species known to reside in or migrate through Jamaica Bay. Courtesy of Brooklyn College.

and Able, 1996). The correlation between water quality and fish community remains vaguely known.

Complicating efforts to understand resilience of the Jamaica Bay fish community is that many species are part of broadly distributed and highly mobile marine populations that move in and out of the bay seasonally. These include popular game fish such as summer and winter flounder, bluefish, and striped bass, all of which use the bay on a seasonal basis. The population dynamics of these species are likely controlled at much broader spatial scales. For example, summer flounder are generally considered to be a single population from Cape Hatteras, North Carolina, to the northern extent of their range in New England (Jones and Quattro, 1999). Fluctuations in the abundance of larval summer flounder in coastal bays have been shown to be related to the abundance of this entire population (Able et al., 2011). Consequently, changes in the abundance of marine fish such as summer flounder in Jamaica Bay are likely to reflect processes (e.g., fishing and predation) in the coastal environment to an equal or greater extent than processes within the bay. Strong connections between oceanic and estuarine fish communities are reported for the similar Barnegat Bay estuary (Able, 2005).

The importance of Jamaica Bay for fish habitat should also be considered in light of climate change and shifting fish distributions. For example, juvenile summer flounder are rarely found in estuaries north of New Jersey (Kraus and Musick, 2001). However, the distribution of summer flounder has been shifting northward (Pinsky et al., 2013). The long-term resilience of the population will depend on the future availability of nursery habitats not yet used by this species, including, perhaps, Jamaica Bay.

Marine Mammals

Jamaica Bay and nearby waters provide important habitat for migratory marine mammals, including pinnipeds and cetaceans. Although no formal survey of marine mammals in Jamaica Bay has been completed, humpback whales (*Megaptera novaeangliae*) occasionally feed in the harbor just outside the inlet, and common dolphin (*Delphinus delphis*), bottlenose dolphin (*Tursiops truncatus*), and sperm whale (*Physeter macrocephalus*) have been found stranded in the bay (USFWS, 1997). Harbor seals (*Phoca vitulina*) are probably the most common marine mammal in Jamaica Bay. Harbor seal populations along the eastern United States have been stable or increasing (Gilbert et al., 2005); there are increased sightings of harbor seals in waters around New York City. From September to May, the shoreline, docks, and jetties of Jamaica Bay (e.g., Breezy Point) serve as essential haul-out sites for this species. As generalists, harbor seals will consume a wide variety of fish, crustaceans, and cephalopods throughout the water column (Payne and Selzer, 1989), suggesting that the bay may be a rich foraging ground for these predators.

Fringing Coastal Vegetation

Perimeter areas of the bay have historically followed a tight zonation, from fringing marsh (low marsh, high marsh, and salt-shrub) (figure 5-3) to upland natural vegetation such as coastal shrub, then maritime forest or maritime grassland (figure 5-4) (Edinger et al., 2008). Within each vegetation type, stressors have changed many of the original formations into degraded types (Edinger et al., 2002; Handel et al., 2013). The Jamaica Bay district has been documented as containing more than 450 plant species (Stalter and Greller, 1988; Stalter and Lamont, 2002), including 12 rare and endangered plant species (Stalter et al., 1996). Many woody and herbaceous nonnative species are abundant. An intensive pilot study of 1.2 miles (2 km) of fringing habitat near Canarsie Pier (Handel et al., 2013) showed the wide extent of invasive species that dominate the shore (bittersweet [*Celastris orbiculatus*], mulberry [*Morus alba*], tree of heaven [*Ailanthus altissima*], Russian olive [*Elaeagnus umbellata*], and others). However, small populations of native fringing vegetation persist. This current, urban mix is typical of the region and other urbanized estuaries (figure 5-5).

Figure 5-3. A salt marsh. Ribbed mussels (*Gukensia demissa*) attach to each other and to salt marsh cordgrass (*Spartina alterniflora*) in this narrow band of low salt marsh fringing Jamaica Bay. High salt marsh plants grow slightly higher in the tidal profile. Inland of the marsh, the beach grades into coastal shrubland, then taller trees. Courtesy of Brooklyn College.

Figure 5-4. Coastal sand communities. Breezy Point's strong salty winds prune the bayberry (*Morella pensylvanica*) of this coastal shrubland. Little bluestem (*Schizachyrium scoparium*) displays autumn seeds in the adjacent coastal grassland. Beaches and dunes, with vegetation communities of different ages such as these, are characteristic of the wind- and wave-tossed natural environments on the sandy south shore of Long Island. Courtesy of Christina Kaunzinger of Rutgers University.

Figure 5-5. Coastal upland forest. Much of the coastal upland is invaded by nonnative species, including phragmites reed (*Phragmites australis*), oriental bittersweet vine (*Celastrus orbiculatus*), and tree of heaven (*Ailanthus altissima*). These competitors often squeeze out or smother native trees and shrubs. Courtesy of Christina Kaunzinger of Rutgers University.

Marsh Islands

Significant marsh loss has occurred since the mid-twentieth century (figure 5-6). The upland between the edge of the bay and the hardscape of the city (figure 5-7) will become even narrower with sea level rise. The reasons for marsh loss are complex, and there is no consensus on which mechanisms are most important. It may be reduction in sediment (Gordon and Houghton, 2004), coupled with increasing wave action from shipping and storms, abetted by sea level rise (Hartig et al., 2002). A study by Deegan et al. (2012) implicated high concentrations of nitrogen in degrading the organic matter "glue" of marsh sediments. Bertness et al. (2014) argued for a cascade from overfishing to overabundance of crabs, which led to direct loss of marsh plants in Narragansett Bay. Analogous effects may be occurring here. The increased levels of phytoplankton due to nitrogen loads may have caused increases in populations of the ribbed mussel, with densities >1,000/square

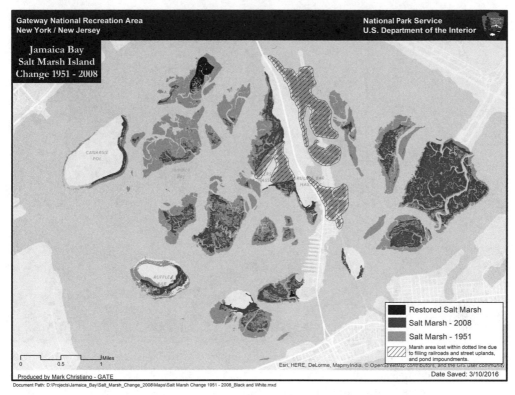

Figure 5-6. Marsh loss and restoration on Jamaica Bay islands, 1951–2008. Courtesy of the Gateway National Recreation Area, National Park Service.

foot (>10,000/m^2) (Franz, 2001). These mussels occur over the entire intertidal zone of salt marshes, with highest densities along marsh edges and creek banks. The extent to which they stabilize marshes (Bertness, 1984) or cause slumping and hasten erosion of marsh edges at high densities (Franz, 2001) is poorly understood.

The most recent losses to the salt marshes tend to occur around their margins. The margins become mudflats, and here the macroalga *Ulva* sp. prospers. Between 1951 and 1974, and 1979 and 1989, the marshes decreased by about 17.8 acres per year (7.2 ha/yr). Between 1994 and 2001, this rate accelerated to 32 acres per year (13 ha/yr) (Hartig et al., 2002). Government agencies and nonprofits have restored more than 150 acres (60 ha) of salt marsh areas to parry these losses (U.S. Army Corps of Engineers, 2016).

Birds

Jamaica Bay is well known for high numbers and species diversity of birds (Kieran, 1982) and is a major destination for the birding community (figure 5-8). At least 325 bird species use the bay's habitats (Trust for Public Land and NYC Audubon, 1987, 1993). The bay and

Figure 5-7. Coastal upland town. Buildings and impervious surfaces occupy the former salt marshes at Coleman Square in Howard Beach on the north shore of Jamaica Bay. Landscaping with coastal grassland, shrub, and forest species will help support native vegetation and wildlife of the bay and improve drainage. This photograph was taken by David Shankbone in June 2007, courtesy of Wikimedia Commons.

surrounding terrestrial environments provide a wide range of habitats, which attract both nesting and migrating species. Before human development, the shorelines were likely used extensively for foraging by herons and egrets, the mudflats were used for feeding by migrant shorebirds, and the islands provided nesting habitat for a range of colonial nesting birds, as well as small song birds. With ditching for mosquito control and the urbanization of the watershed, many of the natural habitats have disappeared or decreased, with the overall effect of decreasing both the diversity and density of birds (Elbin and Tsipoura, 2010). Similarly, the islands in Jamaica Bay have provided important habitat for colonially nesting waterbirds since their resurgence in the city in the 1980s (Bernick, 2007; Brown et al., 2001b; Elbin and Tsipoura, 2010). Freshwater, available at the Jamaica Bay Wildlife Refuge West Pond prior to Hurricane Sandy, provides critical foraging areas for species such as the glossy ibis as they raise their young.

However, threats to avian biodiversity are plentiful. For example, JFK Airport on the eastern bay has no tolerance for birds (Brown et al., 2001a). Plantings that attract birds to the area (e.g., berry-bearing trees and shrubs) are threats not permitted on or near the

Figure 5-8. Great egrets (*Ardea alba*) rest and nest on the interior marsh islands of the bay, while black-crowned night herons stroll by (*Nycticorax nycticorax*). Jamaica Bay is an important place for birds and birdwatchers in New York City. Hundreds of species use the habitats that Jamaica Bay affords. Courtesy of Gateway National Recreation Area, National Park Service.

airport. Additionally, a growing population of raccoons (*Procyon lotor*) (predators) and human disturbance weakens the value of the nesting islands. Data on avian distribution and abundance from the 1980s can serve as a basis for understanding the species that can be present (Burger, 1988).

Reptiles

Diamondback terrapins are the iconic reptiles of Jamaica Bay, often nesting in full view of visitors (figure 5-9). Although terrapins were harvested to virtual extinction for food in the region until the early twentieth century, little was known about terrapins in Jamaica Bay until Cook (1989). Since then, the reproductive biology of this population of terrapins has been the subject of numerous studies. Terrapins are salt marsh specialists, swimming up shallow channels into *Spartina* marshes with the high tide to feed on invertebrates such as snails, clams, and crabs, and back out to deeper water at low tide (Brennessel, 2006). Only adult females come ashore, looking for sunny nest sites in summer. Nearly all nests are quickly predated by raccoons, rats, and other predators (Burger, 1977; Feinberg and Burke, 2003). In some areas nest predators such as raccoons destroy up to 90 percent of nests and hatchlings, and sea level rise will eliminate many safe nesting sites on sandy places around Jamaica Bay islands.

Figure 5-9. A diamondback terrapin (*Malaclemys terrapin*). Terrapins are an iconic species of the Atlantic coast. A salt marsh specialist challenged by habitat loss and pollution, and, around Jamaica Bay, by high populations of raccoons (*Procyon lotor*) that prey on eggs and hatchlings. Courtesy of Ryan Hagerty of the U.S. Fish and Wildlife Service.

Because terrapins have temperature-dependent sex determination (Burke and Calichio, 2014) and most Jamaica Bay nests are warm, most hatchlings are females. Hatchlings quickly find cover (Burger, 1976) and commonly spend their first winter on land (Muldoon and Burke, 2012). There are at least two large Jamaica Bay terrapin populations—one nests on the island of Rulers Bar, the other at JFK Airport. Both are threatened by urbanization, pollution, ongoing salt marsh loss, and high populations of raccoons.

Although diamondback terrapins are very important for Jamaica Bay, the region once had a large and varied herptile fauna. Urbanization resulted in a reduction in diversity and species abundance. Many species were extirpated, and others had their populations severely reduced. From 1980 to 1995, Cook (2008) and others reintroduced many species into Gateway National Recreation Area; several of these were successful (Cook, 2008). This work, and continued monitoring, is extremely important for the recreation area and provides important information for urban ecosystems generally. Species that are present include garter snake, Fowler's toad, spring peepers, grey tree frog, spotted salamander, redbacked salamander, snapping turtle, painted turtle, and box turtle, among others. Of these, the most typical Atlantic coastal species are the diamondback terrapin and Fowler's toad.

Insect Groups of Special Interest

Insect species in urbanized areas have received less attention than in other habitat types. Even in the New York metropolitan area, however, insects have remarkably high biodiversity (Lutz, 1941) and provide a variety of ecosystem services, from pollination to food for other species. The fate of insect ecosystem services is unknown in our changing climate, which is known to cause range shifts of lepidopterans in Europe (Parmesan et al., 1999; Hickling et al., 2005). This pattern may be repeated here for the more than eighty butterfly species known from the area (Cech, 1993). These shifts influence herbivory and mutualism patterns, changing broader community dynamics.

Management decisions can immediately affect insect diversity. For example, the at-risk northeastern beach tiger beetle (*Cicindela dorsalis*) used to occur on New York City beaches but was extirpated from all of the state by the 1950s as a result of change in beach structure, dune stabilization, and damaging human recreation activities. Another beach-dependent species, the hairy-necked tiger beetle (*C. hirticollis*), may be following in its unfortunate footsteps (figure 5-10). It has declined across Long Island and New York City but remains at less developed sites, including beaches of Jamaica Bay (Mawdsley et al., 2013). Like beaches, salt marshes have a dependent insect fauna, including the salt marsh

Figure 5-10. A hairy-necked tiger beetle (*Cicindela hirticollis*). Protecting some beaches from foot and vehicle traffic may enhance resilience of the insect fauna, such as this tiger beetle, and other beach invertebrates in the social-ecological system that is the Jamaica Bay watershed. Courtesy of Matthew Schlesinger of the New York Natural Heritage Program.

tiger beetle (*C. marginata*) and the seaside dragonlet (*Erythrodiplax berenice*), both of which rely on habitats that may be unable to migrate upslope as the sea rises.

With the growth of urban agriculture, native pollinators are of increased public interest. Pollinators are also necessary for reproduction of many coastal plant species. What must be done in management and education to encourage a diverse insect community? Green infrastructure such as green roofs and bioswales is slowly being added, providing connecting habitats that allow pollinators to move through fragmented urban environments.

Coastal fauna can recover from disturbances when their populations are not highly fragmented, but it is unknown whether small, highly fragmented populations can recover as easily. What is certain is that stressors such as habitat loss, climate change, and invasive species cause community changes across trophic levels (Tallamy, 2009).

Needs and Opportunities for Resilience

It is not clear yet what the cumulative effect of recent resilience measures will be. Our knowledge of Jamaica Bay's ecology is still scattershot, with few species or interactions studied in depth or over a long time. Our ecological knowledge of Jamaica Bay resembles a litany of facts and observations, but lacks a holistic understanding of the ecological dynamics of the system, or how those dynamics will respond to resilience intervention measures. Here we highlight several themes to build a more substantive ecological theory of Jamaica Bay as a resilient urbanized watershed.

Linking Physical Forcing Factors to Ecological Conditions

Many of the past declines in Jamaica Bay's ecology have been driven by changes in the physical environment (summarized in chapter 4): paving the watershed, hardening the shore edge, dredging channels, filling marshes, and massive increases in nitrogen inputs. Meanwhile the climate is continuing to raise the sea level. There is an important set of questions about how future changes in physical environment will result in changes in ecological condition. Jamaica Bay represents a local example of threats to marshes and wetlands playing out nationally and internationally. For example, work in Barnegat, Chesapeake, and San Francisco Bays shows analogous issues of urban degradation (Nichols et al., 1986; Boesch et al., 2001; Kennish et al., 2007). Those engaged in the ecosystem science of the bay remain cognizant of research and restoration efforts elsewhere, which inform our activities.

Jamaica Bay's salt marsh islands provide an illustrative example. Their area has declined since at least the mid-twentieth century, but the mechanism of loss remains unclear. As a result, the most cost-effective measures for restoring marshes over the long run are also unclear. We need to monitor marshes, and the concomitant physical factors, to plan better strategies for enhancing them. This understanding will emerge through integrated efforts to understand marsh ecology, biogeochemistry, and hydrodynamics in a resilience context.

A similar set of questions can be asked for mussel and oyster populations. Much relevant work has been done in the Chesapeake (Jackson et al., 2001), and it seems reasonable to expect that changes in circulation and sediment dynamics have affected larval supply and recruitment in Jamaica Bay as well. Mussels in turn may play a role in consolidating marsh sediments and in accretion rates, but we do not understand how important those roles are in stabilizing marshes with respect to other factors. These questions, focused on drivers and particular groups of organisms, point to the need for a robust ecosystem model for Jamaica Bay (see chapter 8). Such a model would begin as a conceptual framework and working hypothesis of major interrelationships of organisms and environmental factors.

Systematically Charting Species Distributions and Abundance

We lack long-term monitoring records for most species in Jamaica Bay, limiting the ability for managers to understand the abundance of species and habitat zones and to target management action to increase species and numbers, as discussed in chapter 7. For example, how do different fish species use Jamaica Bay? Which species are foraging and reproducing, and how important are habitats such as salt marshes as nursery grounds? We need to better understand how marine species are connected to nearshore marine environments outside the Rockaway Inlet, and to the larger Hudson River–Raritan Estuary. In nearby Barnegat Bay, the invertebrate community remains biodiverse and abundant, despite urbanized conditions (Taghon, 2015). Similar persistence of marine invertebrates may hold true in Jamaica Bay or water quality stressors may have altered their resilience.

Similarly, it would be helpful to understand how plant populations in marshes and upland areas vary by jurisdiction, land use history, and management treatment. This interaction has been shown in San Francisco Bay (Nichols, 2016). Many native species are in decline, and introduced species are nearly ubiquitous. What is the best use of effort for plant species management?

Mapping and Measuring Ecosystem Services

People often cite the ecosystem goods and services provided by Jamaica Bay, as we have done. But what are these exactly? How many of these goods and how much of what services are provided? Categories of services have been codified (TEEB, 2011), but we need to measure Jamaica Bay's ecosystem services, understand precisely their dynamics across time and space, and how interventions—whether to restore ecosystems or harden shorelines—will change goods and services. Several of the computer models (see chapter 8), such as the Visionmaker model, provide a basis for estimating ecosystem services.

A related need is that, although we presume these goods and services are valuable, we must articulate them in terms that are broadly appreciated across the Jamaica Bay communities. In part that is a matter of measurement and mapping, but it is also about valuation. Because most decisions about Jamaica Bay are tied to economic considerations, future support for the bay's ecological system will depend on communicating the economic values of the lands and waters of the bay.

Integration from Land to Water, Built to Natural, Jamaica Bay to Region

A major problem in understanding the ecology of Jamaica Bay has been isolation of research topics. Some scientists study the water. Others study coastal environments, and yet others work in the urbanized watershed. We need to integrate these so that we can better frame how changes in the watershed affect ecological dynamics in the estuarine environment and vice versa. These interplays have been identified in other urban estuaries: Barnegat (Kennish et al., 2007), Chesapeake (Dauer et al., 2000), and San Francisco Bays (Douglass and Pickel, 1999).

We need to integrate efforts across jurisdictional boundaries, because species do not see the invisible lines that government authorities or communities draw. We also need to understand how species use environments, from built landscapes to more natural ones. For example, can bioswales and backyards provide habitat for some species? Are there ways to bring "bay habitats" into the urban environment, protecting communities while also encouraging connection to nature? We need to understand how Jamaica Bay fits in to the larger regional ecology as defined by New York City, the New York seascape, and the Northeast, and how this important stopover in the large-scale movements of migratory birds and fish can be enhanced.

Finally, we note that the broad-based support for Jamaica Bay's biodiversity and wildlife must be translated into action. Aims should be set in a realistic way, recognizing all the many kinds of goals that people have for the watershed. Ecological resilience of the kind that Jamaica Bay experienced prior to development is probably not realistic (Hobbs et al., 2009; Handel, 2013), so restoring resilience to Jamaica Bay is a long-term endeavor that requires changes in the watershed and the region as it does in Jamaica Bay proper.

References

Able, K.W. 2005. A re-examination of fish estuarine dependence: Evidence for connectivity between estuarine and ocean habitats. *Estuarine, Coastal and Shelf Science* 64, no. 1: 5–17.

Able, K.W., Sullivan, M.C., Hare, J.A., Bath-Martin, G., Taylor, J.C., and Hagan, R. 2011. Larval abundance of summer flounder (*Paralichthys dentatus*) as a measure of recruitment and stock status. *Fishery Bulletin* 109, no. 1: 68.

Beck, M.W., et al. 2003. The role of nearshore ecosystems as fish and shellfish nurseries.

Issues in Ecology 11.

Benotti, M.J., Abbene, M., and Terracciano, S.A. 2007. *Nitrogen Loading in Jamaica Bay, Long Island, New York: Predevelopment to 2005.* U.S. Geological Survey Scientific Investigations Report 2007–5051. Accessed at: http://pubs.usgs.gov/sir/2007/5051/.

Bernick, A.J. 2007. *New York City Audubon's Harbor Herons Project: 2007 Nesting Survey.* New York City Audubon. Accessed at: http://nycaudubon.org/pdf/2007_NYCA_HH_Report_Bernick-1.pdf.

Bertness, M.D. 1984. Ribbed mussels and *Spartina alterniflora* production in a New England salt marsh. *Ecology*: 65(6): 1794–1807.

Bertness, M.D. 2006. *Atlantic Shorelines: Natural History and Ecology.* Princeton, NJ: Princeton University Press.

Bertness, M.D., Brisson, C.P., Bevil, M.C., and Crotty, S.M. 2014. Herbivory drives the spread of salt marsh die-off. *PloS One* 9, no. 3: e92916.

Black, F.R. 1981. *Jamaica Bay: A History.* Study No. 3. Division of Cultural Resources, North Atlantic Regional Office, National Park Service, U.S. Department of the Interior, Washington, D.C.

Boesch, D.F., Brinsfield, R.B., and Magnien, R.E. 2001. Chesapeake Bay eutrophication. *Journal of Environmental Quality* 30, no. 2: 303–320.

Botton, M.L., and Loveland, R.E. 2011. Temporal and spatial patterns of organic carbon are linked to egg deposition by beach spawning horseshoe crabs (*Limulus polyphemus*). *Hydrobiologia* 658, no. 1: 77–85.

Botton, M.L., Loveland, R.E., and Tiwari, A. 2003. Distribution, abundance, and survivorship of young-of-the-year in a commercially exploited population of horseshoe crabs *Limulus polyphemus*. *Marine Ecology Progress Series* 265: 175–184.

Botton, M.L., Colón, C.P., Rowden, J., Elbin, S., and Sclafani, M. 2014. Impacts of a beach nourishment project on spawning horseshoe crabs in an urban estuary impacted by Hurricane Sandy. Paper presented at the 43rd Annual Benthic Ecology Meeting, Jacksonville, Florida.

Brennessel, B. 2006. *Diamonds in the Marsh: A Natural History of the Diamondback Terrapin.* Hanover, NH: University Press of New England.

Briski, E., Ghabooli, S., Bailey, S.A., and MacIsaac, H.J. 2012. Invasion risk posed by macroinvertebrates transported in ships' ballast tanks. *Biological Invasions* 14, no. 9: 1843–1850.

Brown, K.M., Erwin, R.M., Richmond, M.E., Buckley, P.A., Tanacredi, J.T., and Avrin, D. 2001a. Managing birds and controlling aircraft in the Kennedy Airport–Jamaica Bay Wildlife Refuge Complex: The need for hard data and soft opinions. *Environmental Management* 28, no. 2: 207–224.

Brown, K.M., Tims, J.L., Erwin, R.M., and Richmond, M.E. 2001b. Changes in the nesting populations of colonial waterbirds in Jamaica Bay Wildlife Refuge, New York, 1974–1998. *Northeastern Naturalist* 8: 275–292.

Burger, J. 1976. Behavior of hatchling diamondback terrapins (*Malaclemys terrapin*) in the field. *Copeia*: 1976(4):742–748.

Burger, J. 1977. Determinants of hatching success in diamondback terrapin, *Malaclemys terrapin. American Midland Naturalist* 97, no. 2: 444–464.

Burger, J. 1988. Jamaica Bay studies VIII: An overview of abiotic factors affecting several avian groups. *Journal of Coastal Research* 4: 193–205.

Burke, R.L., and Calichio, A.M. 2014. Temperature-dependent sex determination in the diamond-backed terrapin (*Malaclemys terrapin*). *Journal of Herpetology* 48, no. 4: 466–470.

Cantwell, A.E., and diZerega Wall, D. 2001. *Unearthing Gotham: The Archaeology of New York City*. New Haven: Yale University Press.

Cech, R. 1993. *A Distributional Checklist of the Butterflies and Skippers of the New York City Area (50-mile radius) and Long Island*. New York Butterfly Club.

Clemants, S.E., and Moore, G. 2005. The changing flora of the New York metropolitan region. *Urban Habitats* 3, no. 1: 192–210.

Colón, C.P., and Rowden, J. 2014. Blurring the roles of scientist and activist through citizen science. In: *Civic Learning and Teaching*, Finley, A. (ed). 45–52. Washington, D.C.: Bringing Theory to Practice.

Cook, R.P. 1989. A natural history of the diamondback terrapin. *Underwater Naturalist* 18, no. 1: 25–31.

Cook, R.P. 2008. Potential and limitations of herpetofaunal restoration in an urban landscape. In: *Urban Herpetology*, Mitchell, J.C., Jung Brown, R.E., and Bartholomew, B. (eds). 465–478. Salt Lake City, Utah: Society for the Study of Amphibians and Reptiles.

Dauer, D.M., Ranasinghe, J.A., and Weisberg, S.B. 2000. Relationships between benthic community condition, water quality, sediment quality, nutrient loads, and land use patterns in Chesapeake Bay. *Estuaries* 23, no. 1: 80–96.

Deegan, L.A., Johnson, D., Warren, R.S., Peterson, B.J., Fleeger, J.W., et al. 2012. Coastal eutrophication as a driver of salt marsh loss. *Nature* 490, no. 7420: 388–392.

Douglass, S.L., and Pickel, B.H. 1999. The tide doesn't go out anymore—the effect of bulkheads on urban bay shorelines. *Shore and Beach* 67, no. 2: 19–25.

Edinger, G.J., Evans, D.J., Gebauer, S., Howard, T.G., Hunt, D.M., and Olivero, A.M. 2002. *Ecological Communities of New York State*. Albany: New York State Department of Environmental Conservation.

Edinger, G.J., Feldmann, A.L., Howard, T.G., Schmid, J.J., Eastman, E., et al. 2008. Vegetation Classification and Mapping at Gateway National Recreation Area. Technical Report NPS/NER/NRTR-2008/107. National Park Service. Northeast Region. Philadelphia, PA.

Elbin, S.B., and Tsipoura, N.K., editors. 2010. The Harbor Herons Conservation Plan, New York/New Jersey Harbor Region. New York/New Jersey Harbor Estuary Program. Accessed at: http://www.harborestuary.org/reports/harborheron/ConservationPlan_for_HarborHerons0610.pdf.

Engelhart, S.E., and Horton, B.P. 2012. Holocene sea level database for the Atlantic coast of the United States. *Quaternary Science Reviews* 54: 12–25.

Faurby, S., King, T.L., Obst, M., Hallerman, E.M., Pertoldi, C., and Funch, P. 2010. Population dynamics of American horseshoe crabs—historic climatic events and recent anthropogenic pressures. *Molecular Ecology* 19, no. 15: 3088–3100.

Feinberg, J.A., and Burke, R.L. 2003. Nesting ecology and predation of diamondback terrapins, *Malaclemys terrapin,* at Gateway National Recreation Area, New York. *Journal of Herpetology* 37, no. 3: 517–526.

Franz, D.R. 1982. An historical perspective on mollusks in Lower New York Harbor, with emphasis on oysters. In: *Ecological Stress and the New York Bight: Science and Management*, Meyer, G.F. (ed). 181–192. Columbia, SC: Estuarine Research Federation.

Franz, D.R. 2001. Recruitment, survivorship, and age structure of a New York ribbed mussel population *(Geukensia demissa)* in relation to shore level—a nine-year study. *Estuaries* 24, no. 3: 319–327.

Franz, D.R., and Harris, W.H. 1998. Seasonal and spatial variability in macrobenthos communities in Jamaica Bay, New York—an urban estuary. *Estuaries* 11, no. 1: 15–28.

Gilbert, J.R., Waring, G.T., Wynne, K.M., and Guldager, N. 2005. Changes in abundance of harbor seals in Maine, 1981–2001. *Marine Mammal Science* 21, no. 3: 519–535.

Gillings, S., Atkinson, P.W., Bardsley, S.L., Clark, N.A., Love, S.E., et al. 2007. Shorebird predation of horseshoe crab eggs in Delaware Bay: Species contrasts and availability constraints. *Journal of Animal Ecology* 76, no. 3: 503–514.

Gordon, A.L., and Houghton, R.W. 2004. The waters of Jamaica Bay: Impact on sediment budget. In: *Proceedings, Jamaica Bay's Disappearing Salt Marshes*, New York Sea Grant, Ed., 18. New York: Jamaica Bay Institute, Gateway National Recreational Area, National Park Service.

Grosholz, E.D., and Ruiz, G.M. 1996. Predicting the impact of introduced marine species: Lessons from the multiple invasions of the European green crab *Carcinus maenas*. *Biological Conservation* 78, no. 1: 59–66.

Handel, S.N. 2013. Shelter from the storm. *Ecological Restoration* 31(4): 345–346.

Handel, S.N., Kaunzinger, C.M.K., Johnson, L.R., Corrigan, K.P., and Young, T. 2013. *Shore Parkway, Brooklyn Ecological Assessment*. Report to the City of New York Department of Parks and Recreation.

Hartig, E.K., Gornitz, V., Kolker, A., Mushacke, F., and Fallon, D. 2002. Anthropogenic and climate-change impacts on salt marshes of Jamaica Bay, New York City. *Wetlands* 22, no. 1: 71–89.

Hendrick, D.M. 2006. *Images of America: Jamaica Bay*. Charleston, South Carolina: Arcadia Publishing.

Hickling, R., Roy, D.B., Hill, J.K., and Thomas, C.D. 2005. A northward shift of range margins in British Odonata. *Global Change Biology* 11, no. 3: 502–506.

Hobbs, R.J., Higgs, E., and Harris, J.A. 2009. Novel ecosystems: Implications for conservation and restoration. *Trends in Ecology & Evolution* 24, no. 11: 599–605.

Hoellein, T.J., and Zarnoch, C.B. 2014. Effect of eastern oysters (*Crassostrea virginica*) on sediment carbon and nitrogen dynamics in an urban estuary. *Ecological Applications* 24, no. 2: 271–286.

Horton, R., Bader, D., Kushnir, Y., Little, C., Blake, R., and Rosenzweig, C. 2015a. New York City Panel on Climate Change 2015 Report Chapter 1: Climate Observations and Projections. *Annals of the New York Academy of Sciences* 1336: 18–35. doi:10.1111/nyas.12586.

Horton, R, Little, C., Gornitz, V., Bader, D., and Oppenheimer, M. 2015b. New York City Panel on Climate Change 2015 Report Chapter 2: Sea Level Rise and Coastal Storms. *Annals of the New York Academy of Sciences* 1336: 36–44. doi:10.1111/nyas.12593.

Jackson, J.B.C., Kirby, M.X., Berger, W.H., Bjorndal, K.A., Botsford, L.W., et al. 2001. Historical overfishing and the recent collapse of coastal ecosystems. *Science* 293, no. 5530: 629–637.

Jackson, N.L., Nordstrom, K.F., and Smith, D.R. 2010. Armoring of estuarine shorelines and implications for horseshoe crabs on developed shorelines in Delaware Bay. In: *Puget Sound Shorelines and the Impacts of Armoring—Proceedings of a State of the Science Workshop May 2009*, Shipman, H., Dethier, M.N., Gelfenbaum, G., Fresh,

K.L., and Dinicola, R.S. (eds). 195–202. U.S. Geological Survey Scientific Investigations Report 2010–5254, Accessed at: http://pubs.usgs.gov/sir/2010/5254/pdf/sir20105254.pdf.

Jones, W.J., and Quattro, J.M. 1999. Genetic structure of summer flounder (*Paralichthys dentatus*) populations north and south of Cape Hatteras. *Marine Biology* 133, no. 1: 129–135.

Jordan, T.E., and Valiela, I. 1982. A nitrogen budget of the ribbed mussel, *Geukensia demissa,* and its significance in nitrogen flow in a New England salt marsh. *Limnology and Oceanography* 27, no. 1: 75–90.

Kennish, M.J., Bricker, S.B., Dennison, W.C., Glibert, P.M, Livingston, R.J., et al. 2007. Barnegat Bay-Little Egg Harbor Estuary: Case study of a highly eutrophic coastal bay system. *Ecological Applications* 17, no. sp5: S3–S16.

Kieran, J. 1982. *A Natural History of New York City: A Personal Report after Fifty Years of Study and Enjoyment of Wildlife within the Boundaries of Greater New York.* New York: Fordham University Press.

Kraemer, G.P, Sellberg, M., Gordon, A., and Main, J. 2007. Eight-year record of *Hemigrapsus sanguineus* (Asian shore crab) invasion in western Long Island sound estuary. *Northeastern Naturalist* 14, no. 2: 207–224.

Kraus, R.T., and Musick, J.A. 2001. A brief interpretation of summer flounder, *Paralichthys dentatus,* movements and stock structure with new tagging data on juveniles. *Marine Fisheries Review* 63, no. 3: 1–6.

Kreeger, D.A., and Newell, R.I.E. 2001. Seasonal utilization of different seston carbon sources by the ribbed mussel, *Geukensia demissa* (Dillwyn) in a mid-Atlantic salt marsh. *Journal of Experimental Marine Biology and Ecology* 260, no. 1: 71–91.

Leschen, A.S., and Correia, S.J. 2010. Mortality in female horseshoe crabs *(Limulus polyphemus)* from biomedical bleeding and handling: Implications for fisheries management. *Marine and Freshwater Behaviour and Physiology* 43, no. 2: 135–147.

Levinton, J., Doall, M., and Allam, B. 2013. Growth and mortality patterns of the eastern oyster *Crassostrea virginica* in impacted waters in coastal waters in New York, USA. *Journal of Shellfish Research* 32, no. 2: 417–427.

Lutz, F.E. 1941. *A Lot of Insects: Entomology in a Suburban Garden.* New York: G.P. Putnam's Sons.

Mattei, J.H., Beekey, M.A., Rudman, A., and Woronik, A. 2010. Reproductive behavior in horseshoe crabs: Does density matter? *Current Zoology* 56, no. 5: 634–642.

Mawdsley, J.R., Schlesinger, M.D., Simmons, T., and Blanchard, O.J. 2013. Status of the tiger beetle *Cicindela hirticollis* Say (Coleoptera: Cicindelidae) in New York City and on Long Island, New York, USA. *Insecta Mundi.* Paper 822. Accessed at: http://digitalcommons.unl.edu/insectamundi/822.

McHugh, C.M., Hartin, C.A., Mountain, G.S., and Gould, H.M. 2010. The role of glacioeustasy in sequence formation: Mid-Atlantic Continental Margin, USA. *Marine Geology* 277, no. 1: 31–47.

Miller, B. 2000. *Fat of the Land: Garbage in New York: The Last Two Hundred Years.* New York: Four Walls Eight Windows.

Muldoon, K.A., and Burke, R.L. 2012. Movements, overwintering, and mortality of hatchling diamond-backed terrapins (*Malaclemys terrapin*) at Jamaica Bay, New York. *Canadian Journal of Zoology* 90, no. 5: 651–662.

New York City Department of Environmental Protection. 2007. *Jamaica Bay Watershed*

Protection Plan. Accessed at: http://www.nyc.gov/html/dep/html/harborwater/jamaica_bay.shtml.

New York City Department of Environmental Protection. 2009. *New York City's Wastewater Treatment System.* Accessed at: http://www.nyc.gov/html/dep/pdf/wwsystem.pdf.

New York City Department of Environmental Protection. 2014. *Jamaica Bay Watershed Protection Plan 2014 Update.* Accessed at: http://www.nyc.gov/html/dep/pdf/jamaica_bay/jbwpp_update_10012014.pdf.

Nichols, F.H. 2016. *The San Francisco Bay and Delta—An Estuary Undergoing Change.* Accessed at: http://sfbay.wr.usgs.gov/general_factsheets/change.html. Accessed February 28, 2016.

Nichols, F.H., Cloern, J.E., Luoma. S.N., and Peterson, D.H. 1986. The modification of an estuary. *Science* 231: 567–573.

Parmesan, C., Ryrholm, N., Stefanescu, C., Hill, J.K., Thomas, C.D., et al. 1999. Poleward shifts in geographical ranges of butterfly species associated with regional warming. *Nature* 399, no. 6736: 579–583.

Payne, P.M., and Selzer, L.A. 1989. The distribution, abundance and selected prey of the harbor seal, *Phoca vitulina concolor,* in southern New England. *Marine Mammal Science* 5, no. 2: 173–192.

Peteet, D., Lieberman, L., Higgiston, P., Sritraitrat, S., and Kenna, T., 2008. Yameko to JFK—Paleoenvironmental Change from Jamaica Bay Marshes. Presented at the Review of the "State of the Bay" Jamaica Bay Scientific Symposium, City of New York Department of Environmental Protection, New York.

Pinsky, M.L., Worm, B., Fogarty, M.J., Sarmiento, J.L., and Levin, S.A. 2013. Marine taxa track local climate velocities. *Science* 341, no. 6151: 1239–1242.

Robinson, G.R., Yurlina, M.E., and Handel, S.N. 1994. A century of change in the Staten Island flora: Ecological correlates of species losses and invasions. *Bulletin of the Torrey Botanical Club* 121: 119–129.

Sanderson, E.W. 2009. *Mannahatta: A Natural History of New York City.* New York: Abrams.

Sanderson, E.W. (2016). Cartographic evidence for historical geomorphological change and wetland formation in Jamaica Bay, New York. *Northeastern Naturalist* 23(2): 277–304.

Sirkin, L.A. 1967. Late-Pleistocene pollen stratigraphy of western Long Island and eastern Staten Island, New York. *Quaternary Paleoecology* 7: 249–274.

Socrates, J. 2010. A Study of the Striped Bass in the Marine District of New York State: Juvenile Striped Bass. New York State Department of Environmental Conservation, Completion Report AFC-33, 2010.

Stalter, R., Byer, M.D., and Tanacredi, J.T. 1996. Rare and endangered plants at Gateway National Recreation Area: A case for protection of urban natural areas. *Landscape and Urban Planning* 35, no. 1: 41–51.

Stalter, R., and Greller, A. 1988. A floristic inventory of the Gateway National Recreation Area, New York–New Jersey. *Rhodora* 90: 21–26.

Stalter, R., and Lamont, E.E. 2002. Vascular flora of Jamaica Bay Wildlife Refuge, Long Island, New York. *Journal of the Torrey Botanical Society* 129: 346–358.

Swanson, R.L., and Wilson, R.E. 2008. Increased tidal ranges coinciding with Jamaica Bay development contribute to marsh flooding. *Journal of Coastal Research* 24: 1565–1569.

Szedlmayer, S.T., and Able, K.W. 1996. Patterns of seasonal availability and habitat use by fishes and decapod crustaceans in a southern New Jersey estuary. *Estuaries* 19, no. 3: 697–709.

Taghon, G. 2015. Benthic Invertebrate Community Monitoring and Indicator Development for Barnegat Bay-Little Egg Harbor Estuary: Year 2. Plan 9: Research, Project SR12-006. New Jersey Department of Environmental Protection, Office of Research, Trenton, NJ.

Tallamy, D.W. 2009. *Bringing Nature Home: How You Can Sustain Wildlife with Native Plants, Updated and Expanded.* Portland, Oregon: Timber Press.

TEEB—The Economics of Ecosystems and Biodiversity. 2011. *TEEB Manual for Cities: Ecosystem Services in Urban Management.* Accessed at: www.teebweb.org.

Trust for Public Land and the New York City Audubon Society. 1987. *Buffer the Bay: A Survey of Jamaica Bay's Unprotected Open Shoreline and Uplands.* New York: The Trust for Public Land.

Trust for Public Land and the New York City Audubon Society. 1993. *Buffer the Bay Revisited, an Updated Report on Jamaica Bay's Open Shoreline and Uplands.* New York: Capital Cities/ABC.

U.S. Army Corps of Engineers. 2016. Jamaica Bay Marsh Islands. Accessed at: http://www .nan.usace.army.mil/Missions/CivilWorks/ProjectsinNewYork/EldersPointJamaica BaySaltMarshIslands.aspx.

U.S. Fish and Wildlife Service. 1997. Significant habitats and habitat complexes of the New York Bight watershed: Jamaica Bay and Breezy Point. Complex #16. In: Significant Habitats and Habitat Complexes of the New York Bight Watershed. U.S. Fish and Wildlife Service, Southern New England—New York Bight Coastal Ecosystems Program, Charlestown, R.I. Accessed at: http://nctc.fws.gov/resources/knowledge -resources/pubs5/web_link/text/toc.htm#16.

Waldman, J. 2008. Research Opportunities in the Natural and Social Sciences at the Jamaica Bay Unit of Gateway National Recreation Area. Technical Report, National Park Service, Gateway National Recreation Area, Jamaica Bay Institute, Brooklyn, New York.

Wallace, R.B. and Gobler, C.J. 2015. Factors controlling blooms of microalgae and macroalgae (*Ulva rigida*) in a eutrophic, urban estuary: Jamaica Bay, NY, USA. *Estuaries and Coasts* 38: 519–533.

Walls, E.A., Berkson, J., and Smith, S.A. 2002. The horseshoe crab, *Limulus polyphemus:* 200 million years of existence, 100 years of study. *Reviews in Fisheries Science* 10, no. 1: 39–73.

Wroth, L.C. 1970. *The Voyages of Giovanni da Verrazzano, 1524–1528.* New Haven, Connecticut: Yale University Press.

Zarnoch, C.B., and Schreibman, M.P. 2012. Growth and reproduction of eastern oysters, *Crassostrea virginica,* in a New York City estuary: Implications for restoration. *Urban Habitats* 7.

6

Neighborhood and Community Perspectives of Resilience in the Jamaica Bay Watershed

Laxmi Ramasubramanian, Mike Menser, Erin Rieser,
Mia Brezin, Leah Feder, Racquel Forrester, Shorna Allred,
Gretchen Ferenz, Jennifer Bolstad, Walter Meyer,
and Keith Tidball

The challenges and opportunities for resilience for urban estuaries such as Jamaica Bay come from the people who live, work, and visit there. Soils, rocks, and the weather; birds, fish, and salt marsh grasses, may or may not be resilient on their own terms, but what human beings do and how we think is where resilience practice by people begins. Throughout this book, we refer to the Jamaica Bay watershed as a social-ecological system (SES), and in this chapter we attempt to describe the varied populations, many neighborhoods, and diverse communities of this part of New York City and to explore how they think of themselves and their relationship to the environment.

Between March and August 2014, we conducted thirty-three interviews with key stakeholders and community leaders and nine focus groups with a total of forty-five participants. Discussion in the focus groups and interviews was guided by the following topics: (1) best practices in terms of social and economic resilience; (2) existing community resources; (3) community needs for building and maintaining social, economic, infrastructural, and environmental resilience; (4) how communities weather disruptions, with a particular focus on Hurricane Sandy in 2012; and (5) personal, recreational, and community relationships to Jamaica Bay. As noted in chapter 1, many of these neighborhoods were hard hit by the flooding (figure 6-1). Data were subsequently analyzed within several organizing themes that reflect the challenges

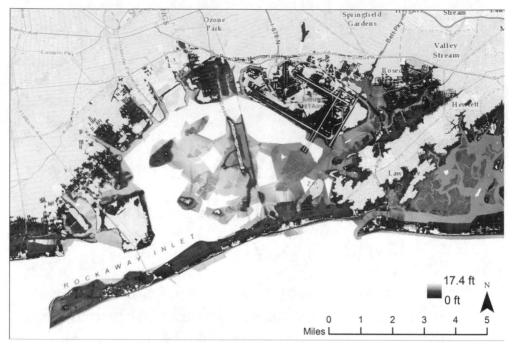

Figure 6-1. Extent of flooding in Jamaica Bay communities caused by the storm surge associated with Hurricane Sandy. Data from Federal Emergency Management Agency Modeling Task Force (MOTF)-Hurricane Sandy Impact Analysis available at https://www.arcgis.com/home/item.ht ml?id=307dd522499d4a44a33d7296a5da5ea0. The base map is from Esri, HERE, DeLorme, MapmyIndia, © OpenStreetMap contributors, and the GIS user community. Courtesy of Mario Giampieri of the Wildlife Conservation Society.

and opportunities for developing best practices for building resilience in Jamaica Bay neighborhoods. Similar work, focusing on visitors to the parklands around Jamaica Bay, has been conducted by the U.S. Forest Service through the New York City Urban Field Station and should be read as a complement to our work (Campbell et al., 2014; Campbell et al., 2015a, b).

We begin by highlighting basic demographic aspects describing the Jamaica Bay watershed today, and then turn to a discussion of neighborhood and community resilience. The notion of community resilience in this chapter and chapter 11 (which describes strategies to enhance community resilience) is primarily based on the views of Jamaica Bay residents themselves, as revealed through our interviews and a review of published research about the impacts of Hurricane Sandy on Jamaica Bay and its surrounding neighborhoods. Unlike the perspectives taken in chapters 4 and 5, which provide historical context, the communities are very much focused on recent events, in particular Hurricane Sandy (as described in chapter 1.) We conclude with some thoughts on research needs regarding community resilience.

Diverse Demographics of Jamaica Bay

The names of the communities around Jamaica Bay are redolent with the idea of being close to the sea: Sheepshead Bay, Gerritsen Beach, Edgemere, Broad Channel, Far Rockaway and Breezy Point (figures 6-2 through figure 6-7). Although spatial proximity to Jamaica Bay is shared among all groups and neighborhoods surveyed, the diversity of relationships to Jamaica Bay largely rests around historic, economic, and cultural factors.

Figure 6-2. Nostrand Houses in Sheepshead Bay, Brooklyn, looking south across Batchelder Street and Avenue Y. Photograph by Jim Henderson in June 2012, courtesy of Wikimedia Commons.

Examining political jurisdictions, the Jamaica Bay watershed covers seven community boards within New York City, including four in Brooklyn (community boards 5, 13, 15, and 18) and three in Queens (10, 13, and 14). It also includes a small portion of the Town of Hempstead in Nassau County, over the New York City boundary, that was not surveyed for this chapter, but which does compose approximately 20 percent of the topographic watershed of Jamaica Bay (color plate VI).

Population density is significantly higher in the Brooklyn neighborhoods, consistently between approximately 20,000–55,000 people per square mile, while in the Queens neighborhoods that were surveyed, most census tracts were below 30,000 people per square mile, with the exception of 41,000 people per square mile in one tract in Far Rockaway, where public housing is located. These higher densities are in stark contrast to

Figure 6-3. Houses and docks line the banks of Gerritsen Creek in Gerritsen Beach, Brooklyn, not far from Floyd Bennett Field and the Rockaway Inlet. Photograph by GK tramrunner229 in November 2006, courtesy of Wikimedia Commons.

the population density of around 2,000 people per square mile on the western end of the peninsula at Breezy Point (color plate VIII).

According to the American Community Survey's five-year estimates (2006–2010), average household incomes in the watershed using 2010 inflation-adjusted dollars ranged from approximately $23,000 per year (e.g., particular census tracts in Far Rockaway, Rockaway Peninsula, Queens) to $168,000 (e.g., specific census tracts in Neponsit and Belle Harbor, also on the Rockaway Peninsula, Queens). Census data from 2010 suggest that higher household incomes are more closely associated with demographically white communities, while communities on the lower end of the economic spectrum tend to be more ethnically and racially diverse (NYC Special Initiative on Recovery and Resiliency, 2013).

Again, relying on 2010 census data, racial and ethnic demographics vary considerably between and within the Jamaica Bay watershed communities, often over the distance of a few miles. Some communities are nearly all white (95–100 percent), including Belle Harbor, Neponsit, Breezy Point, Gerritsen Beach, and Broad Channel. Others, especially bordering the northwest and far southeastern areas of Jamaica Bay, including Canarsie,

Figure 6-4. Public housing, like these New York City Housing Authority buildings in Cypress Hills, Queens, is an important aspect of the residential neighborhoods surrounding Jamaica Bay. Photograph by Jim Henderson in May 2009, courtesy of Wikimedia Commons.

Rockaway Park, and Far Rockaway, are composed of between 50 and 90 percent African Americans, with Hispanic groups making up 15–30 percent of the population.

These demographic differences are important because they closely correlate with divergent understandings and perspectives about neighborhood and community resilience, pre– and post–Hurricane Sandy in 2012. We observed throughout the course of our interviews that these socio-demographic differences were often mentioned as playing a role in distinguishing a given community's relationship to Jamaica Bay and were used by community members to describe a community's ability to be socially and economically resilient in the wake of environmental and other disruptions. For example, more spatially dense communities often spoke more of needing better community leadership and access to resources such as public space, job training, and youth education/recreational programs. Meanwhile, communities that had less density per square mile spoke more about repairing and/or maintaining the value of private property and environmental protection of Jamaica Bay, as was the case in Breezy Point and Broad Channel. Such communities felt they had established strong ties between neighbors that made their communities resilient. Again, the census data do not readily explain distinctions that individuals make among the different factors that make a particular community resilient, but rather help to contextualize neighborhood-level differences within broader historical, social, economic, infrastructural, and geographic contexts within which these perspectives about community resilience are forged.

Figure 6-5. Springfield Lake, located within Springfield Park in the Springfield Gardens neighborhood of Queens, is a reminder of Thurston's Creek, which once flowed through this area, across what is now John F. Kennedy International Airport, and out to Jamaica Bay. Photograph by Peter Greenburg in February 2013, courtesy of Wikimedia Commons.

Rich Network of Institutions in Jamaica Bay

There is a rich network of governmental organizations that work in Jamaica Bay. Some governmental agencies are working on ecosystem restoration in Jamaica Bay (for example, the New York City Department of Environmental Protection, the New York State Department of Environmental Conservation, the National Park Service, and the U.S. Fish and Wildlife Service). Other agencies focus on monitoring the health of the bay (New York City Department of Environmental Protection, New York State Department of Environmental Conservation, U.S. Army Corps of Engineers, the National Park Service, New York City Natural Resources Group, U.S. Environmental Protection Agency, and the U.S. Geological Survey). There are also those with responsibilities for holistic planning of areas in and around the bay (New York City Department of City Planning). Agencies that have legal authority over other aspects of the SES include the Port Authority (for their work with John F. Kennedy International Airport) and the National Oceanic and Atmospheric Administration. There are also a number of nongovernmental organizations working in Jamaica Bay, such as the American Littoral Society, Jamaica Bay Ecowatchers, and the new Science and Resilience Institute at Jamaica Bay, to name a few that have a distinct focus on protecting and maintaining social-ecological resilience of the watershed. In addition,

Figure 6-6. Duplex in the Far Rockaway neighborhood of Queens. Far Rockaway is part of the barrier island system on the south side of Long Island that separates Jamaica Bay from the Atlantic Ocean. Photograph by David Shankbone in August 2013, courtesy of Wikimedia Commons.

there are several hundred community-based organizations that address different aspects of social services provision, such as housing, economic development, education, health-care, and so on. After Hurricane Sandy, these community-based organizations became involved in the larger conversations about community resilience in Jamaica Bay.

Issues of Community Resilience in Jamaica Bay Neighborhoods

Throughout the interviews we conducted with individuals and organizations in Jamaica Bay, several issues about community resilience, as it related to the geography, physical infrastructure, and social and political relationships in Jamaica Bay communities came up repeatedly. These can be summarized under the ten themes listed in table 6-1 and used to organize the discussion below.

Accessibility and Mobility

People living near Jamaica Bay often discussed their feelings of being "cut off" from the rest of New York City. In addition to the distance (10–25 miles) that must be traversed to reach the central business districts in Manhattan and Brooklyn, there are limited routes

Table 6-1. Factors related to community resilience.

Accessibility and mobility
Isolation as benefit
Infrastructure
Housing
Commercial development
Employment
Youth engagement and education
Local understanding of risk and vulnerability
Community resources, social networks, and communication
Relationship to governments and nonprofits

Table 6-1. Factors related to community resilience discovered through interviews with residents of the Jamaica Bay watershed.

for entering and exiting some Jamaica Bay communities via car. The subway (A train), the AirTrain shuttle, the Long Island Rail Road, and buses provide public transit options, but residents considered these options to be "limited." A temporary ferry service was started after Hurricane Sandy, but it ended in October 2014 (Norris, 2014).

Within the context of discussing Hurricane Sandy, numerous participants emphasized the limited routes for evacuation from their residential locations. This was most acute on Rockaway Peninsula, which has few connections to other parts of Brooklyn, Queens, and Nassau County. The Cross Bay Boulevard and the Gil Hodges Memorial Bridge are the main north-south arteries for entering the peninsula, and traffic congestion is a constant problem. Primary routes for entering and exiting the peninsula by car were identified as the Marine Parkway Bridge into south Brooklyn, the Cross Bay Bridge into Queens, and Route 878 through Nassau County (Long Island).

Respondents in many of our focus groups indicated that the limited ingress and egress points for entering and leaving the peninsula, particularly during evacuation, were a factor that explained why certain people stayed during the storm despite the many warnings. During the Far Rockaway focus group, this point was re-emphasized, as several participants described that the longer folks waited to evacuate, the harder it became to leave the peninsula, as bridges were closed (the closer the storm came to shore) and major thruways out of Far Rockaway (through Nassau County) were choked with heavy traffic. Furthermore, those who decided to evacuate later, either just before or during the storm,

Figure 6-7. Looking northeast from St. Virgilius Church at a volunteer fire station on a mostly sunny afternoon in Broad Channel, a community on an island in the center of Jamaica Bay. Photograph by Jim Henderson in April 2010, courtesy of Wikimedia Commons.

found bridges closed and, eventually, the roads were flooded with water and debris. For those reliant on public transportation, these routes were also shut down in the hours leading up to the storm due to the citywide public transit closure, leaving many with no option but to shelter in place. Due to storm damage, it was seven months before the subway and shuttle trains returned to their pre-Sandy levels of service (Flegenheimer, 2013).

The inability of the Metropolitan Transit Authority to reinstate service on the Rockaway Peninsula had a devastating effect on many groups, especially those who rely on public transportation as their primary mode of transit. As numerous focus group participants emphasized, those on the lower end of the socioeconomic spectrum regularly use the bus or train as their primary mode of transit and are less likely to afford private vehicles. Unfortunately, they are also more likely to have a significant distance to traverse on their regular journey to work and therefore are in greater need of a reliable transportation infrastructure. Many residents routinely take two or more modes of transportation. Transportation system disruptions further complicated their commute and added time to their travel schedule. The poorest of the poor remain within their communities, in large part due to the lack of easy, efficient, and affordable transportation options.

Many focus group participants assured us that all residents, even those who owned or had access to cars, recognized how underserved almost all the neighborhoods surrounding Jamaica Bay are by public transportation, especially when compared to many other parts of New York City. Although bus service is adequate in several neighborhoods, every Jamaica Bay neighborhood resident faces the same challenges of needing to make multiple transfers to get from their residence to anywhere else in the city—be it for work, recreation, school, or to access other public services.

In the aftermath of Hurricane Sandy, some of those who relocated lost jobs because of untenable commute times and costs. Others were able to remain (were "resilient") to the extent that they were able to stay in their homes but ended up losing their jobs because Hurricane Sandy temporarily deprived them of viable transportation alternatives (flooded cars, disrupted mass transit options). Without a job, some individuals eventually lost their housing.

Although accessibility and mobility were probably a challenge before Hurricane Sandy, the catastrophic event exacerbated disparities between different individuals and neighborhoods. During our interviews, we realized that the hurricane and its aftermath shaped the public's understanding of resilience.

Isolation as Benefit

As discussed, inaccessibility, for some communities, creates a sense of isolation. However, our focus groups and conversations revealed that isolation is sometimes viewed as a positive attribute. Numerous interviewees noted that either their personal or family's decision to move into the area—and in some cases their decision to purchase second homes in Jamaica Bay neighborhoods—reflected desires to live in safer communities, be closer to the water, and to live in proximity to, but not directly in, the city. This was particularly important to the "island communities" that are usually no more than a few blocks from the water—Breezy Point, Gerritsen Beach, Broad Channel—and are smaller geographically, more economically advantaged, and racially homogenous than other communities surveyed in Jamaica Bay. The desire to live in a quieter, smaller, and more isolated community was also coupled with a sense of neighborliness that accompanies living in a small community where everyone knows, and cares about, everyone else. The uniqueness of these communities, which was universally remarked upon within the focus groups, is partially a result of this inaccessibility.

Although living in isolated communities that are more vulnerable to fluctuations in the environment posed significant threats to these communities, their geographic isolation and insularity corresponded to social resilience. These communities consistently stated that the community networks, leadership, and "culture of volunteerism" cultivated in these smaller communities enable them to be better prepared to manage social and environmental disruptions, allocating resources and coordinating communication efficiently and effectively.

Many residents mentioned the high percentage of city public service employees living within their communities—fire, police, and sanitation—suggesting that their access to city services is potentially superior to other neighborhoods in Jamaica Bay despite their isolation. These city workers provide built-in, if informal, connections to city government. Interestingly though, on the one hand, these communities are isolated geographically, which poses a risk during disturbance events, and on the other, their isolation simultaneously increases community resilience and facilitates opportunities for neighbors to work together. It is important to recognize the significance of individual choice and sense of control that seems highly correlated to viewing living in isolation as a benefit.

Infrastructure

When we discussed issues of infrastructure with Jamaica Bay community residents, they identified a number of roads, including the Belt Parkway and the Cross Bay Boulevard, that serve as east-west and north-south arteries to connect the region; bridges such as the Gil Hodges Memorial Bridge; trains, including the A train and the Long Island Rail Road; the many bus routes; a wastewater treatment plant; and the natural infrastructure of the area's beaches as important to resilience. "Hard" infrastructure categories were frequently cited in resident focus groups as a crucial aspect of maintaining a resilient community, as well as helping to build back after the destruction of Hurricane Sandy. Alongside concerns with "hard infrastructure" was a consistent concern about housing, especially the different impacts Hurricane Sandy had on homeowners and renters.

Throughout the Jamaica Bay region, regular flooding from rain and high water tables is common. Several residents acknowledge that this is in part due to their neighborhoods being built on "sand and swamplands," which they understand makes their properties and infrastructure more vulnerable to inundation from water and storms. Restoring infrastructure related to storm drains and sewers is high on the current agenda in Canarsie, and local leaders have made repairing the sewage system a priority. The Rockaway Peninsula's sewer system is being built out and improved following extensive damage from Hurricane Sandy, but reports of sinkholes from damaged pipes were common in many of our conversations with local residents. The general incapacity of local storm drain systems to withstand heavy rainfall was a concern for many residents, particularly those who are homeowners, even prior to Hurricane Sandy. The Rockaway Wastewater Treatment Plant, for example, was constructed in a low-lying area to facilitate drainage of sewer pipes, but this position makes it vulnerable to damage from coastal flooding.

Some residents recognize that natural landscape provides a degree of storm protection, especially for waterfront communities on the Rockaway Peninsula; however, as resident participants in our focus groups pointed out, they believe that beaches need to be further buttressed with dunes, groins, berms, and other forms of coastal protection. Dunes were particularly effective at protecting coastal communities during Hurricane Sandy: Long Beach, on Long Island, which chose not to build dunes, suffered around $200 million in damages, while neighboring dune-protected Bradley Beach had only approximately $3 million in damages (Navarro and Nuwer, 2012).

Residents suggested that a variety of forms of coastal protection be implemented, including both "hard" and "soft," or "green," infrastructure (see chapter 9). Residents mentioned flood gates, a dome, sandbags, dunes, elevating entire communities or homes, and reestablishing the marshes. Residents whose homes are particularly close to Jamaica Bay— Broad Channel and parts of Far Rockaway, in particular—emphasized the need to regularly maintain bulkheads as a crucial aspect of protecting their homes and communities from storm surges. As a Far Rockaway resident said, "That does not have to happen. This is the parks area; how could they not know? They know how to make nature and society cooperate together. Why aren't they using sand to protect these areas? Building dunes . . . how we can use sand as a deterrent . . . how to have these dunes work for us [to protect us]."

Many residents spoke of investing their own time and money into trying to maintain bulkheads, acquiring sandbags, and other forms of mitigation to reduce the impact of storm surges and flooding on their property. Others spoke about using their own resources and taking the time to fix those kinds of infrastructure nearest their homes, but indicated that this spot-fix approach was less than desirable and not adequate in the long run. "Out of devastation they had to put up their own money to fix things . . . people who have their own money could afford their own relief," stated a Canarsie focus group participant. Several residents felt that coastal protection infrastructure should be more regularly maintained by the city and not by individual homeowners.

Housing

Homeownership rates in Jamaica Bay neighborhoods vary dramatically: In Far Rockaway only 26 percent of the population are homeowners, whereas in the neighborhoods of Breezy Point and Howard Beach, 95 percent of the population own homes (NYC Special Initiative for Rebuilding and Resiliency, 2013). We found that renters and homeowners have different relationships to place that are pertinent for resilience practice. For example, since Hurricane Sandy, homeowners have been eligible for a range of assistance, from federal, state, and city programs, as well as from privately purchased homeowners' insurance policies. Although homeowners expressed frustration with these resources and how they have been deployed, they were generally aware of their existence and how to access them, in contrast to renters, who had many fewer programs that were accessible and useful to serve their needs.

Prior to Hurricane Sandy, since 2008, housing foreclosures have been a persistent type of disturbance. The issue of housing insecurity was most pronounced in the Canarsie focus group, but was also mentioned in Gerritsen Beach and on the Rockaways. Foreclosures were amplified by Hurricane Sandy, in particular where some homeowners were illegally renting out their basement apartments to generate income to help pay mortgages. Many of these basement apartments were flooded and subsequently vacated by tenants, leaving their owners to bear the costs of repair alongside loss of rent. For many homeowners in the Jamaica Bay region, the bulk of the owners' wealth is invested in their homes, which meant that they had little access to the necessary capital to rebuild their homes in the wake of major structural damage such as that experienced during Hurricane Sandy, outside of government programs. Many individuals opined that New York City's failure to adequately structure and disperse recovery and rebuilding funds further exacerbated the precarious situation of many people.

In contrast to homeowners, many renters were displaced from their homes altogether, lacking any recourse to stay in place. A renter in Rockaway Park said, "Renters didn't get anything. If you owned a house, then you get insurance. People with houses still have money coming. $30K and I'm sitting here with nothing." In Arverne there were concerns about rents increasing due to the costs of repairing and elevating homes: "Landlords are going to be required to elevate as well, [there is] going to be a tremendous cost to them—passed on to renters." Programs and resources targeted specifically to renters' needs might increase community resilience by allowing renters to remain within their homes and in turn promote neighborhood stability through enhanced economic activity.

Commercial Development

The push for job training centers, job creation, and better educational programming for youth was a consistent call across communities on all sides of Jamaica Bay. Education and employment are key to community resilience and are often thought of before environmental and infrastructure concerns or even the need for strong community and local governmental leadership.

In many communities throughout Jamaica Bay, particularly those that consist of mixed residential and commercial or industrial areas, residents expressed frustration with the lack of retail diversity and insufficient presence of services and amenities. Where new businesses have opened, there has been high turnover rate, which residents attributed to a lack of access to capital, leading to an inability for new small businesses to sustain themselves. Where development has occurred in recent years, there have been, in some instances, concerns over access to new community resources and questions of which residents are being served. The seemingly unequal access to commercial resources—stores, groceries, and restaurants—also fueled tension among and between communities within

Jamaica Bay, particularly along class and racial lines. Certain communities that are demo-graphically whiter and wealthier, such as Broad Channel or Breezy Point, are predomi-nantly residential and generally do not appear to favor commercial development. How-ever, based on our conversations, there is a prevalent opinion that these communities could have more regular access to these opportunities (if they chose to), and this was a source of consternation for communities that felt underserved. Part of the reason for such discrepancies could be correlated to population density; however, this fact was not explic-itly cited by participants and interviewees but rather implied in their insistence about the lack of resources (and the need for them) in their respective communities.

Employment

The economic hardships facing some Jamaica Bay communities are also expressed in the high unemployment rate in some areas. In the Rockaways, in particular, residents attrib-uted high unemployment to changes that occurred in two of the largest employers on the Rockaways: St. John's Episcopal Hospital (closed) and Madelaine Chocolate (down-sized following damage due to Hurricane Sandy). It was generally felt there was a lack of employment opportunities on the Rockaway Peninsula. Although not necessarily a prominent concern for part-time residents of the peninsula or those who rely on their own personal forms of transit to get to employment outside of the peninsula, many of the residents on the eastern parts of the peninsula find geographic isolation, poor schools, and a lack of sufficient training and opportunities for employment to be a daily hardship. Based on interviews we conducted with community organizers, such as those individuals working with Alliance for a Just Rebuilding, a coalition of labor unions and faith-based, community, environmental, and policy organizations working to address equity in Sandy recovery and rebuilding, we learned that most residents are forced to travel for hours to acquire basic training required for many blue-collar jobs, representing a significant time and cost burden.

It is clear that disturbance events such as Sandy have exacerbated economic struggles, both directly and indirectly. For those directly affected by Hurricane Sandy through flood-ing, the storm created financial hardship due to the need for significant up-front financial outlays, increased travel costs for displaced residents, lost income, and decreases in home value. As numerous residents made clear, however, things were already quite dire in many communities before the storm, where high unemployment, poor public schools, and per-ceived disinvestment on behalf of the city had poised several communities to be "hit even harder" by the disruption of Hurricane Sandy. The notion that places such as Rockaway Park or Far Rockaway were "already disasters" before the storm hit, as mentioned in inter-views, reflects the understanding of the status quo. Residents consistently called for the creation of more local jobs, adequate job training, and the securing of these opportunities

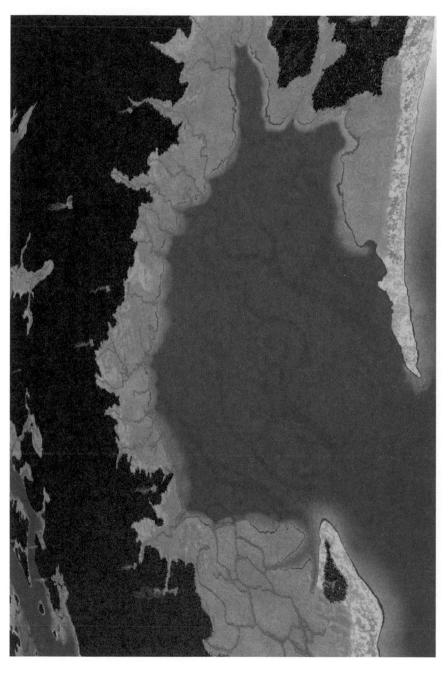

Color plate 1. Jamaica Bay, c. 1609. When Henry Hudson sailed into what would someday be known as Lower New York Harbor in September 1609, he saw "three great Rivers." The northernmost of the three may have been Jamaica Bay, shown here as a wide open lagoon, fringed with salt marshes and extensive upland forests. Lenape settlements are indicated by slight clearings and smoke. Mannahatta (Manhattan) is visible in the distance. Courtesy of Mario Giampieri of The Welikia Project (welikia.org) at the Wildlife Conservation Society.

Color plate II. Jamaica Sound, c. 1679. This colorful detail from "Long Island Sirvaide by Robart Ryder" from approximately 1679 shows the view of Jamaica Bay on the first land-based survey of Long Island. Ryder probably lived in Gravesend, Brooklyn, to the west of Jamaica Bay, which is shown here as "Jamaica Sound," a round open bay on the western end of the island. The "Rookoway" (i.e., Rockaway) spit extends across the southeast side of the sound with a shoal in the mouth. Courtesy of John Carter Brown Library, Brown University.

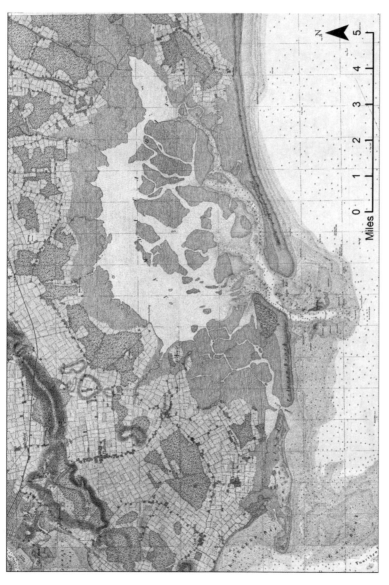

Color plate III. Jamaica Bay, c. 1844. This map, a detail from the "Map of New-York Bay And Harbor And The Environs" prepared under the supervision of Ferdinand Rudolf Hassler of the U.S. Coast Survey and published posthumously in 1844, shows the rural character of the watershed, mainly farm fields and small towns in the mid-nineteenth century. Within the bay are numerous salt marsh islands clustered toward the eastern side, protected by the Rockaway Neck. Extensive salt marshes also fringe the edge of the bay. The bathymetry traces the sediments around the mouth of the inlet and moving east to west with the longshore current on the south side of Long Island. Courtesy of the David Rumsey Map Collection (www.davidrumsey.com), with georeferencing and scale bar by Christopher Spagnoli of The Welikia Project (welikia.org) at the Wildlife Conservation Society.

Color plate IV. Concept for Jamaica Bay improvement, c. 1912. In 1907 the Jamaica Bay Improvement Commission proposed a plan to create a new industrial port to provide "sufficient wharfage at reasonable prices to meet the ever increasing demands of commerce." Although the plan was never fully realized, the project did lead to dredging, landfill, and construction of new bulkheads and other "improvements" that have had a long-lasting effect. The drawing shown is from the cover of a commemorative dinner menu to celebrate the project at Arion Hotel, Rockaway Beach, in 1912. Courtesy of the New York Public Library (digitalcollections.nypl.org/items/2a419c30-f46d-012f-de12-58d385a7bbd0).

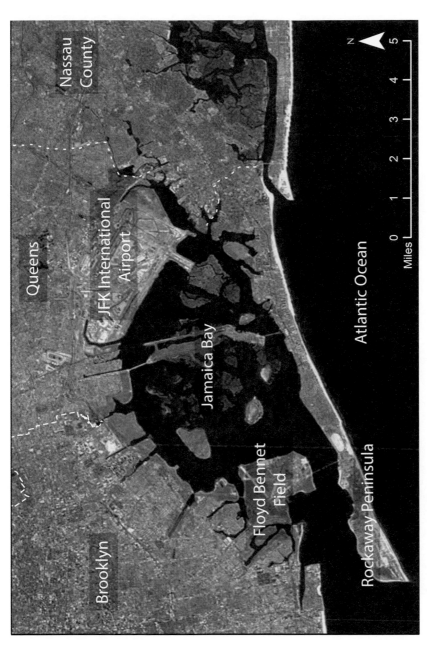

Color plate V. Jamaica Bay, c. 2014. This composite of aerial photography shows the Jamaica Bay social-ecological system (SES) as it appeared around the time of this book's publication. Green colors indicate vegetation, gray colors indicate built infrastructure, and black indicates water. Labels are provided for key features mentioned in the text. Courtesy of Mario Giampieri of the Wildlife Conservation Society based on a base map provided by Esri, DigitalGlobe, GeoEye, Earthstar Geographies, CNES/Airbus DS, USDA, USGS, AEX, Getmapping, Aerogrid, IGN, IGP, swisstopo, and the GIS User Community.

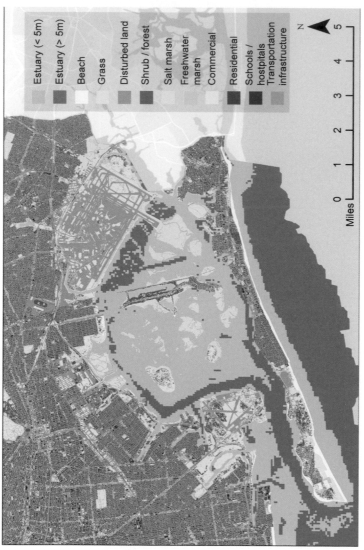

Color plate VI. Map of land cover and water depths of the Jamaica Bay watershed. This map shows the fine-grained nature of the urban watershed with a composite of data sets from approximately 2010 to 2014. Residences, commercial uses, utilities, transportation, and infrastructure stack in close proximity and are organized into rectilinear patterns. Grassy areas fill in between the infrastructure at John F. Kennedy International Airport (JFK), Floyd Bennett Field (a former airport), and on closed and capped landfills. In the waters of the bay, a deeper navigation channel from the Rockaway Inlet to JFK and the "borrow pits" dredged to create the airport sites are visible. The map is courtesy of the Visionmaker Project (visionmaker.nyc) at the Wildlife Conservation Society based on the City of New York's Department of City Planning's MapPLUTO and LION data sets (http://www1.nyc.gov/site/planning/data-maps/open-data/pluto-mappluto-archive.page) and ecological cover mapping from the Natural Areas Conservancy (naturalareasnyc.org).

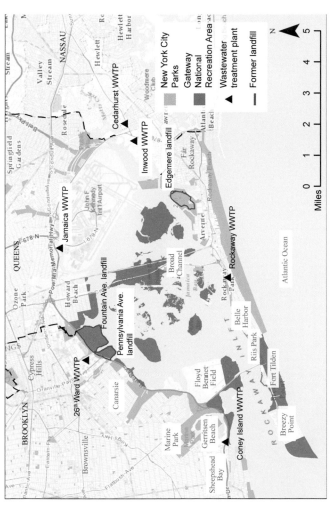

Color plate VII. Map of selected important jurisdictions in the Jamaica Bay watershed. The National Park Service operates several units of the Gateway National Recreation Area, including the Jamaica Bay Wildlife Refuge, Floyd Bennett Field, and Riis Park; New York City Parks manages several adjacent properties west of the Queens/Nassau County line. These federal and city parklands are co-managed under an agreement signed in 2012. Six wastewater treatment plants release treated effluent into Jamaica Bay, four managed by New York City and two by Nassau County. Significant progress has been made in recent decades in lowering nitrogen levels in treated wastewater released from these facilities. There are also three, now closed, landfills on former salt marshes along the margins of the bay. The map is courtesy of Mario Giampieri of the Wildlife Conservation Society based on data from the City of New York's Department of City Planning's MapPLUTO, parks, and landfill data sets (nycopendata.socrata. com) and the New York State Pollution Elimination. Base map details are from some combination of Esri, HERE, DeLorme, Intermap, increment P Corp., GEBCO, USGS, FAO, NPS, NRCAN, GeoBase, IGN, Kadaster NL, Ordnance Survey, Esri Japan, METI, Esri China (Hong Kong), swisstopo, MapmyIndia, © OpenStreetMap contributors, and the GIS user community.

Color plate VIII. Map of population density and race in Jamaica Bay communities. This map detail shows where people of different races live in the communities near Jamaica Bay based on the 2010 U.S. Census. Each dot equals one person, so the density of the dots also gives a sense of relative population density. Courtesy of Weldon Cooper Center for Public Service, Rector and Visitors of the University of Virginia (Dustin A. Cable, creator).

not only in the wake of the rebuilding efforts surrounding the aftermath of Hurricane Sandy but to make such opportunities the norm. This reflects residents' aspirations and desires not only for themselves but for their families and their families' futures.

Youth Engagement and Education

In lower income communities such as Rockaway Park, Far Rockaway, and Canarsie, focus group participants reported that youth idleness was an issue and emphasized the need for after-school youth programming. According to residents, as well as community leaders, a general sense of alienation from Jamaica Bay, in particular, plus a dearth of education about the ecosystem, combine to form a disconnect between potential resources and their use. In low-income communities of color, many young people do not know how to swim and have little interaction with or knowledge of Jamaica Bay. Some organizations, including the Rockaway Waterfront Alliance, Rockaway Youth Task Force, and You Are Never Alone, are actively working to address these issues in the central and eastern parts of the Rockaway Peninsula. Most youth organizations, however, are underfunded and not sufficiently networked with one another or the larger constellation of available resources to be sufficiently impactful.

In contrast, in more affluent communities, proximity to Jamaica Bay waters was mentioned as a central component of what made the community a positive place to live and grow up for young people, offering opportunities to stay busy and connect with nature through school programming and organizations such as the American Littoral Society and the Jamaica Bay Wildlife Refuge. Many of these organizations are still in operation today and yet their scope seems limited, as many residents from areas outside of the Broad Channel community do not know these organizations exist. Commensurate with the challenges mentioned above, connecting with public schools and New York City Parks Department was cited as often challenging for lower income communities of color, who have to travel a longer distance to reach education facilities on Jamaica Bay. Due to the costs of funding field trips for students, as well as bureaucratic challenges for nonprofit groups to work with the Department of Education to provide in-school programming for youth, these resources remain inaccessible to many groups. Several community leaders indicated that science classes and youth education programs were very desirable.

Local Understanding of Risk and Vulnerability

In relation to environmental risk, many of the communities living in proximity to Jamaica Bay that have experienced regular high-tide flooding tended to downplay the risk related to global climate change and sea level rise. This is perhaps most notable in communities such as Broad Channel where, despite the fact that Hurricane Sandy caused significant damage, many residents feel fairly resilient and capable of dealing with the economic, infrastructural, and social risks that come with living in the middle of Jamaica Bay. Participants in the Broad Channel group noted: "[Our] community handled [the storm] much better than other places because they are a tight-knit community. . . . Food trucks came in every day. There was food and water. I have 50 toothbrushes now. Clothing was piled high." The greatest challenges for Broad Channel after the storm included cleanup, along with the logistics of getting children to the schools to which they had been temporarily reassigned. On possible long-term emotional impacts, one participant noted, "It's like stubbing your toe . . . some people can make it a big deal, most people just keep going." To be sure, many folks in the community did not, and could not, come back or build back. For those who continue to stay, participants stated, some believe that Hurricane Sandy was a unique storm that "probably won't happen again in their lifetime." Nevertheless, the storm was also seen as a wake-up call to take environmental issues, such as sea level rise, more seriously.

For other communities, such as Gerritsen Beach, people were surprised about the wide extent of the damage resulting from Hurricane Sandy. Several residents stated that their homes had never flooded before. Furthermore, focus group participants suggested that Gerritsen Beach residents not only did not know how to prepare to save their property, homes, and cars, but also did not have a cohesive evacuation plan. "We did not know [how bad the flooding would be] and that's why we didn't leave," noted one resident. Many in this group, in particular, underscored the psychological damage of living through the storm, citing it as a long-term effect that is still an issue in their community two years later. The focus group mentioned how mental health services, which continued to be frequented by some residents, provided help for neighbors and children with post-traumatic stress disorder from the storm and that concerns are triggered by subsequent storms.

Perceptions of vulnerability extend to proximity to the water, but also to a sense of security in terms of property and general quality of life in the small neighborhood. As several residents related, they experienced perceived increased theft and insecurity in their neighborhoods in the days and weeks after Hurricane Sandy. The sense of "outsiders" taking advantage of neighbors and friends in desperate and vulnerable situations has shaken up long-standing perceptions of their communities being safe, albeit rather insular, places where everyone knows everyone else and takes care of one another. Similar concerns were echoed in the Broad Channel focus group, although it should be noted that while some participants echoed this feeling, others suggested that this was not true. Whether true or false, it certainly seems that the feeling of vulnerability extends beyond concerns over flooding.

In Far Rockaway and Canarsie, the experiences of severe flooding, power outages, and damage to personal and public property during Hurricane Sandy was equally unexpected.

Participants in focus groups in Canarsie noted that they sometimes run into their basement every time it rains now to see if there is flooding and that it is stressful. Others in the focus groups noted how there has been an "uptick in crime" in their neighborhoods, although not everyone agreed. Some expressed the thought that because communities never imagined that a disturbance event such as Hurricane Sandy *could* happen, they tend to envision themselves as much more vulnerable in the aftermath. As participants in the Far Rockaway focus group noted, "A lot of people did not understand the sense of urgency—because last year the hurricane was fine, so they wanted to stay. Homeowners wanted to stay with their property." Even in more self-defined "waterfront" communities such as Breezy Point, residents felt that the full potential of the storm was not understood, nor was it well communicated to residents, despite the fact that their own community security forces went door to door telling everyone to evacuate.

During interviews, Hurricane Sandy was a recent memory and everyone understood the potential for serious disaster and acute risk for isolated communities. However, there is great variance in terms of a sense about whether such events will happen again in the near future. Some described Sandy as a "once-in-a-lifetime storm," while others expressed knowledge of rising sea levels and climate change. Residents in Broad Channel, who "live by the tide," acknowledged that "25 years ago we didn't have 7-foot tides. That's what we have tonight" [the evening of the focus group]. Some expressed the idea that the new "normal" translates into paying more attention to where they park their cars at night, and making a more considered effort to replant marshlands.

Vulnerability to disturbance was expressed in financial terms for many residents, especially homeowners. Many emphasized how the entirety of a family's savings was invested in homes and cars that were significantly damaged—if not outright destroyed—by Hurricane Sandy. The lack of funds required to pay up front for damages, hotels, or other places stay in the months after the storm fueled much of the criticism of the Build It Back program, New York City's program for helping homeowners to fund the rebuilding of their homes (with federal grants) in the wake of Hurricane Sandy. Build It Back has been criticized for being an "unending loop of lost documents, aborted meetings and frustrating exchanges with temporary workers handling . . . application[s]" and the application process has been called "over designed and undermanaged" (Buettner and Chen, 2014). More than 20,000 homeowners applied for assistance; and as of November 2014 the program had fully repaired only 208 homes, started construction on another 848, and sent out 1,449 reimbursement checks. However, in October 2015, the Mayor's Office of Housing Recovery Operations reported that they had sent a total of 5,272 reimbursement checks totaling $103 million, a rapid increase in the pace of recovery. However, these benefits went only to single-family homeowners and co-op and condominium owners (NYC Mayor's Office of Recovery and Resiliency, 2015).

Community Resources, Social Networks, and Communication

Although in some communities, residents seemed confident about where to go for community information and news in normal times, in others, residents described the lack of community space and shared information as an acute need—often, the most pressing one.

Some of the most commonly used meeting spaces and community hubs around Jamaica Bay are houses of worship. These are also the primary source of information for many residents. Other frequently used spaces include ad hoc or organized community spaces, such as You Are Never Alone, in Rockaway Park, a drop-in community resource center that became a relief hub during Sandy. Gerritsen Beach residents noted that they were lucky enough to have a de facto community space through the volunteer firefighters, and Broad Channel mentioned the Veterans of Foreign Wars and American Legion facilities that provided a vital site for organizing after Hurricane Sandy. Park recreation centers, public libraries, and other public facilities might also be used. Such spaces are often vital to the community, yet exist precariously due to a lack of consistent funding. Nonprofits and community groups active around the bay often operate with small and unreliable budgets, creating uncertainty about their presence in the future. The need for grant writing and fundraising assistance for these types of organizations was emphasized in stakeholder interviews and focus groups.

Many residents cited a need for better communications among community organizations and between community organizations and residents. On the social level, this manifests as a sense of community and mutual care. Participants in some neighborhood focus groups, especially in lower income and under-resourced communities, described confusion about where to go for resources. In these communities, some participants observed infighting among community groups and a sense that residents did not always look out for each other, and noted the long history of insufficient access to resources that had led to a sense of scarcity and mistrust.

Residents pointed to digital social networks as a useful method for reaching younger people, but noted the generational divide: among residents over forty, in particular in low-income communities, social media use around Jamaica Bay drops off significantly. Therefore, more traditional, analog communication modalities should be used in trying to reach older and more vulnerable residents, including flyering at local libraries and around the community, getting notices into a local newspaper (such as the *Rockaway Wave* or *Canarsie Courier*), or communicating with local houses of worship.

For communicating official information, community board meetings and police precinct meetings are useful conduits, although some residents expressed a sense that these meetings were informational as opposed to genuinely participatory. There is also a disparity of citizen involvement in community boards; some communities are very active, while others use different channels for communication and civic engagement. Canarsie participants referred to their local community board 18 as a "joke" and "corrupt," citing primarily the lack of interventions that are made on the behalf of residents by the community board. More strongly coalesced communities had well-established communications infrastructure, including, in Breezy Point, a phone alert system managed by the local cooperative. Despite this, all communities expressed a sense that information could flow better. For instance, residents in Breezy Point expressed a desire for a registry of socially vulnerable residents, while in Canarsie, there is an effort under way to organize block associations that would facilitate communication.

Relationship to Governments (Local, Federal) and Nonprofits

All communities expressed some degree of mistrust of federal and city government, reflecting a history of feeling neglected to the extent that their needs are not on the agenda of government bodies or were not properly represented. It is interesting to note that all the communities expressed this distrust: socioeconomic level and geographic location did not matter. This finding speaks to a sense that resources and the paths to access them are ineffective and, in cases of more marginalized communities, completely opaque. Lack of trust in government has caused some communities to turn inward and become more dependent on local institutions and organizations, along with their neighbors. Many expressed dissatisfaction with New York City's and the American Red Cross's management of Hurricane Sandy recovery and, in the years following, the Build It Back program.

Residents of more marginalized communities traced their mistrust back to decades of neglect by government officials, starting with public housing residents being forced to live in isolated areas, such as Far Rockaway. These communities feel that the government overlooks their needs while focusing on better-off, nearby neighborhoods.

Wealthier communities with more access to resources also reported being mistrustful of government. Focus group participants from these neighborhoods felt that the government viewed them as self-sufficient and able to take care of themselves. As a result, residents in these neighborhoods felt that they did not receive their fair share of recovery assistance or financial aid. Many communities also felt oversurveyed in the wake of Hurricane Sandy, and, in some cases, mistrustful of "experts."

In contrast, residents expressed general satisfaction with the state's New York Rising process, primarily because representatives have been chosen directly from and by communities and because the process has offered communities funding and a direct voice in how to allocate it in their community. Despite the perceived success of this program, most agreed that the neighborhood-by-neighborhood process represented a missed opportunity for crossbay collaboration that would have improved future resilience.

Research Needs for Community Resilience

We are extremely appreciative of the members of the public, as well as representatives from government and nongovernmental organizations, who gave generously of their time to participate in this research and share their views about community resilience with us. One of the major challenges associated with understanding neighborhood and community perspectives of resilience is that these perceptions appear to be dynamic, changing over time. Although this research was conducted eighteen months after Hurricane Sandy, respondents were still affected by their experiences. Their frustrations with the slow pace of recovery, which has since picked up, also affected their views and their responses. For researchers intending to explore the complex meanings of neighborhood and community resilience, we encourage the development of data collection instruments that can capture data quickly and accurately across different slices of time, because we have a sense that attitudes and perceptions of resilience are mutable across time and space. Furthermore, we feel that additional efforts to reach out to hear from the most vulnerable populations are necessary in undertaking research in this area—a goal that is fraught with many challenges but one that we must continue to strive toward to build a stronger, more resilient society.

Acknowledgments: The authors acknowledge Monica Barra, Victoria Curtis, Amanda Lewis, Tim Viltz, and Jeremy Wells for contributing to the interviews in this chapter.

References

Buettner, R., and Chen, D. 2014. Broken Pledges and Bottlenecks Hurt Mayor Bloomberg's Build It Back Effort. *New York Times*. September 4. Accessed at: http://www.nytimes.com/2014/09/05/nyregion/after-hurricane-sandy-a-rebuilding-program-is-hindered-by-its-own-construction.html?_r=0.

Campbell, L., Svendsen, E., Falxa-Raymond, N., Baine, G., 2014. Reading the landscape, a reflection on method. *PLOT* 3: 90–95.

Campbell, L., Svendsen, E., Sonti, N., Johnson, M. 2015a. Reading the Landscape: A Social Assessment of Parks and their Natural Areas in Jamaica Bay Communities. White Paper, Part I: Social Assessment Overview. U.S. Forest Service, New York City Urban Field Station, Queens, NY.

Campbell, L., Svendsen, E., Sonti, N., Johnson, M. 2015b. Reading the Landscape: A Social Assessment of Parks and their Natural Areas in Jamaica Bay Communities. White Paper, Part II: Park Profiles. U.S. Forest Service, New York City Urban Field Station, Queens, NY.

Flegenheimer, M. 2013. Just in Time for Summer, the A Train Is Fully Restored. *New York Times*. May 20. Accessed at: http://www.nytimes.com/2013/05/31/nyregion/the-a-line-returns-to-the-rockaways-mostly-for-the-better.html.

Navarro, M., Nuwer, R., 2012. Resisted for Blocking the View, Dunes Prove They Blunt Storms. *New York Times*. December 3. Accessed at: http://www.nytimes.com/2012/12/04/science/earth/after-hurricane-sandy-dunes-prove-they-blunt-storms

.html?pagewanted=all.

Norris, M. 2014. Rockaway Ferry. *The New Yorker*. October 31. Accessed at: http://www
.newyorker.com/culture/culture-desk/rockaway-ferry.

NYC Mayor's Office of Recovery and Resiliency. 2015. One City, Rebuilding Together,
Progress Update, October 2015. Accessed at: http://www1.nyc.gov/assets/home/
downloads/pdf/reports/2015/One-City-Progress-Report.pdf.

NYC Special Initiative on Recovery and Resiliency. 2013. Chapter 16: South Queens. In:
PlaNYC: A Stronger, More Resilient New York. Accessed at: http://www.nyc.gov/html/
sirr/downloads/pdf/final_report/Ch16_SouthQueens_FINAL_singles.pdf.

PART III

Tools for Resilience Practice

7

Resilience Indicators and Monitoring: An Example of Climate Change Resiliency Indicators for Jamaica Bay

Bernice Rosenzweig, Arnold L. Gordon, John Marra,
Robert Chant, Christopher J. Zappa, and Adam S. Parris

In the first half of the twentieth century, Jamaica Bay underwent a major regime shift—the near complete loss of *Crassostrea virginica*, the American oyster. During the 1800s, Jamaica Bay was famous for the abundance and quality of its oysters. Although historic documents from the seventeenth and eighteenth centuries refer to natural oyster populations and oyster harvesting in Jamaica Bay, the advent of oyster seeding in the 1860s ushered in a robust oystering industry (Black, 1981). "Rockaway" oysters from Jamaica Bay were prized for their size and flavor by restaurateurs, and hundreds of "oystermen" made their living on the bay. The oystermen of 1905—near the peak of the Jamaica Bay oyster industry—likely had no way of knowing that in just thirty-five years not only would the seeded oyster harvesting industry in Jamaica Bay be gone, but that oysters would be virtually absent from Jamaica Bay. They did not have the information necessary to show that Jamaica Bay oysters—and the ecosystem services they provided—were not resilient to the natural and human disturbances of the early twentieth century (figure 7-1).

Even today, the ultimate cause of the demise of the Jamaica Bay oyster is unclear, but the larger point rings true for Jamaica Bay and other urban estuaries—multiple disturbances can act together to change the state of the system. In the case of the oyster, these disturbances likely include the accumulating stress caused by piping poorly treated sewage directly into the bay, increased development in the watershed, and policy decisions such as the cessation of oyster seeding after harvesting was banned due to contamination in 1921. Disturbances also include natural shocks such as the 1938 "Long Island Express" Hurricane, which disrupted remaining shellfish beds in Jamaica Bay. However, without

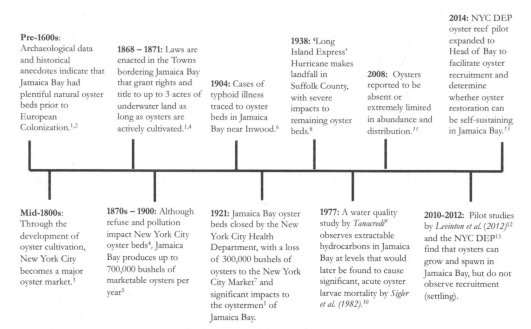

Pre-1600s: Archaeological data and historical anecdotes indicate that Jamaica Bay had plentiful natural oyster beds prior to European Colonization.[1,2]

1868 – 1871: Laws are enacted in the Towns bordering Jamaica Bay that grant rights and title to up to 3 acres of underwater land as long as oysters are actively cultivated.[1,4]

1904: Cases of typhoid illness traced to oyster beds in Jamaica Bay near Inwood.[6]

1938: 'Long Island Express' Hurricane makes landfall in Suffolk County, with severe impacts to remaining oyster beds.[8]

2008: Oysters reported to be absent or extremely limited in abundance and distribution.[11]

2014: NYC DEP oyster reef pilot expanded to Head of Bay to facilitate oyster recruitment and determine whether oyster restoration can be self-sustaining in Jamaica Bay.[13]

Mid-1800s: Through the development of oyster cultivation, New York City becomes a major oyster market.[3]

1870s – 1900: Although refuse and pollution impact New York City oyster beds[4], Jamaica Bay produces up to 700,000 bushels of marketable oysters per year[5]

1921: Jamaica Bay oyster beds closed by the New York City Health Department, with a loss of 300,000 bushels of oysters to the New York City Market[7] and significant impacts to the oystermen[1] of Jamaica Bay.

1977: A water quality study by *Tanacredi*[9] observes extractable hydrocarbons in Jamaica Bay at levels that would later be found to cause significant, acute oyster larvae mortality by *Sigler et al. (1982)*.[10]

2010-2012: Pilot studies by *Levinton et al. (2012)*[12] and the NYC DEP[13] find that oysters can grow and spawn in Jamaica Bay, but do not observe recruitment (settling).

Figure 7-1. Qualitative timeline of oysters in Jamaica Bay. The lack of sustained, quantitative data on oyster populations, water quality, and extreme event impacts makes it impossible to understand the system response to the multiple potential stressors on oyster populations. Courtesy of Bernice Rosenzweig of CUNY Advanced Science Research Center. Data sources: 1) Black, 1981; 2) Hendrick 2006; 3) Mackenzie 1996; 4) Blackford 1884; 5) Franz 1982; 6) Interstate Sanitation Commission 1937; 7) *New York Times* 1921; 8) Grambo and Vega 1984; 9) Tanacredi 1977; 10) Sigler and Liebovitz 1982; 11) Waldman 2008; 12) Levinton et al. 2013; 13) McLaughlin 2014.

consistent information on oyster populations, the ecosystem services they provide and the magnitude of various stresses over time, it is impossible to determine the magnitude to which these disturbances contributed to the collapse of the Jamaica Bay oyster populations. The dramatic loss of the Jamaica Bay oyster provides an example of the need for identifying *resilience indicators*, measurable parameters that track the status, trend, or performance of a social-ecological system (SES) such as Jamaica Bay. Today, with magnified threats to Jamaica Bay from increased watershed development and the imminent stresses and shocks of climate change, it is critical that resilience indicators be identified and a framework developed to monitor them. Such a framework could serve as a model for the many other estuaries facing similar stresses worldwide (Ramesh et al., 2016).

The City of New York is currently developing a climate resiliency indicators and monitoring system to support adaptation planning in New York City (Solecki et al., 2015). Here, we describe the adoption and extension of this approach to better understand the effects of climate change, as well as other human-caused and natural disturbances on Jamaica Bay. We will also discuss the availability of observational data to support resilience analysis and the need for the synthesis of existing data and additional monitoring. The development of a Jamaica Bay climate resilience indicators and monitoring system will provide essential support for comprehensive management for resilience and the long-term evaluation of resilience practices. Such an indicator system will serve as a critical bridge between primary scientific and socioeconomic research and management planning efforts such as the Jamaica Bay Watershed Protection Plan (NYC DEP, 2007) and *OneNYC*, New York City's sustainability plan (City of New York, 2015). Although we will identify key indicators that will be essential for understanding the resilience state of Jamaica Bay and potential impacts of future climate change and management decisions, it is important that local stakeholders and experts collaborate to codesign the full list of resilience indicators for Jamaica Bay or other sites where such a framework is applied.

Preliminary Resilience Indicators for Jamaica Bay

Resilience is difficult to quantify directly, largely as a result of the complexity of SESs (Cumming et al., 2005; see chapter 2 for a detailed discussion of ecological resilience). The New York City Panel on Climate Change (NPCC) (Jacob et al., 2010) established three selection criteria for resilience indicators: policy relevance, analytical soundness, and measurability. To meet each criterion, NPCC provides a set of factors, tailored here to Jamaica Bay:

Policy Relevance

- Provide a representative picture of conditions in the Jamaica Bay SES,
- Measure stakeholder-relevant hazards and society's responses,
- Be simple, easy to interpret, and able to show trends over time,
- Provide a basis for intra- and intercity comparisons,
- Have a scope applicable to critical regional issues, and
- Have a baseline, threshold, or reference value or range of values against which to compare, so users can assess the significance of the values associated with it through time.

Analytical Soundness

- Be theoretically well founded in technical and scientific terms,
- Based on local, national, or international standards with consensus about its validity, and
- Readily linked to economic models, scenario projections, and information systems.

Measurability

- Based on readily available data or data available at a reasonable cost-benefit ratio,
- Be adequately documented and of known quality,
- Be updated at regular intervals, in accordance with reliable procedures, and
- Of sufficient length in time and numbers to allow a quantitative statistical evaluation of the uncertainties associated with the data.

Although the NPCC indicators and monitoring system focuses only on the disturbances caused by global climate change, we extend this approach to consider the many
other long- and short-term disturbances. In the following sections, we discuss initial resilience indicators in light of NPCC's general criteria, using the detailed criteria as a basis
to assess the relevance to decision making and the feasibility for monitoring. For Jamaica
Bay, we can divide the essential resilience indicators into five major categories: (1) *Climate
Hazards*, (2) *Water and Sediment Quality*, (3) *Land Use and Land Cover*, (4) *Biodiversity and
Species Abundance*, and (5) *Community Resilience*. It is critical to understand the interde-
pendencies between the different categories and the role that they play in the Jamaica
Bay SES, advancing our understanding of matter and energy flows among various components and our ability to predict the SES's response to shocks and disturbances. Monitoring
multiple variables simultaneously, at daily or seasonal time scales, enables one to parse
out the relationships critical to resilience (figures 7-2 and 7-3).

Climate Hazard Indicators

Climate hazards from severe storms, coastal and upland flooding, and heat waves are at
the center of the conversation about resilience for Jamaica Bay.

Policy Relevance and Analytical Soundness

Clearly, climate hazards from severe storms and storm surges are of critical relevance to
Jamaica Bay and other coastal watersheds. Climate change is projected to result in further
increases in temperature, precipitation, and extreme events (Walsh et al., 2014; NPCC,
2013), which will have significant impacts on estuarine circulation, sediment transport,
water quality, and biodiversity of Jamaica Bay (Zappa et al., 2003, 2007; MacCready and
Geyer, 2010; Anthony et al., 2009; Orton et al., 2010). From 1900 to 2011, the mean
annual temperature in New York City increased 36.3°F (2.4°C), and the mean yearly rainfall increased by 0.6 feet (19.6 cm) (NPCC, 2013). Sea level at the Battery has risen 1 foot
(0.34 m) since 1900, and evidence suggests it is likely that sea levels may rise faster along
the northeast coast with climate change (NPCC, 2013). These changes will result in an
increased coastal flooding risk, as well as changes in estuarine circulation and retention
times. Indicators of weather and climate for Jamaica Bay will enable managers to characterize the magnitude and frequency of extreme weather events as drivers of disturbance
to Jamaica Bay (e.g., Hurricane Sandy; figure 7-4).

Measurability and Monitoring

Compared with other parts of the New York City metropolitan region, the spatial distribution of meteorological stations in the Jamaica Bay watershed is relatively sparse.
Ongoing weather and climate monitoring is conducted by multiple federal agencies,

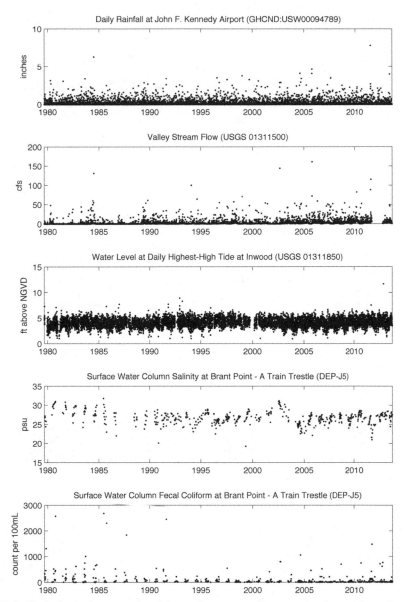

Figure 7-2. Hydrological, oceanographic, and water quality metrics from selected monitoring sites in the northeast section of Jamaica Bay, 1980–2014. Data shown (from top to bottom) include rainfall, streamflow, highest high tide, salinity, and fecal coliform levels. Stakeholder engagement and data analysis are required to convert this kind of monitoring data to indicators of resilience. (cfs-cubic feet per second; psu-practical salinity unit; NGVD = National Geodetic Vertical Datum.) Data from the National Climatic Data Center; National Water Information System (http://waterdata.usgs.gov/nwis/); U.S. Geological Survey New York Water Science Center (Tristen Tagliaferri, pers. comm.); Township of Hempstead Department of Conservation and Waterways (James Browne, pers. comm.); Jamaica Bay Water Quality Database (Brett Branco, pers. comm.). Courtesy of Bernice Rosenzweig of CUNY Advanced Science Research Center.

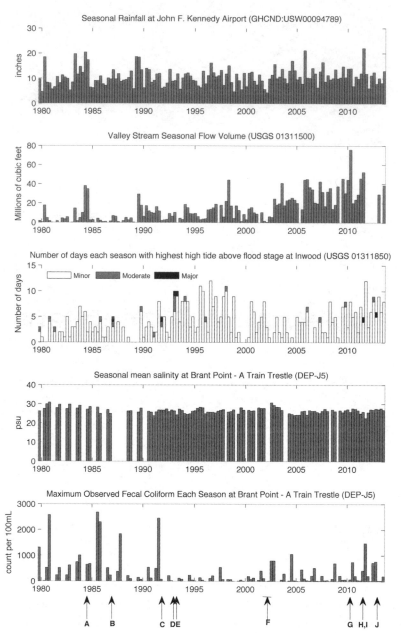

Figure 7-3. Correspondence of extreme weather events and selected hydrological, oceanographic, and water quality metrics averaged on a seasonal basis for Jamaica Bay, 1980–2014. Ten meteorological disturbances shown at bottom: A) June 30, 1984: extreme rain; B) January 2, 1987: nor'easter storm; C) October 30–31, 1991: the "perfect" storm; D) December 10–11, 1992 nor'easter storm; E) March 13–14, 1993: blizzard; F) 2001–2002: drought; G) October 4–5, 2010: extreme rain; H) August 15, 2011: extreme rain; I) August 28, 2011: Hurricane Irene; J) October 29, 2102: Hurricane Sandy. Courtesy of Bernice Rosenzweig of CUNY Advanced Science Research Center.

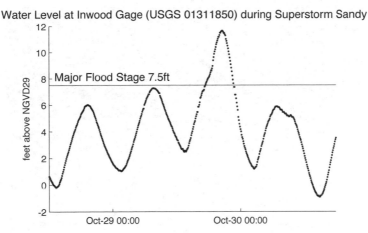

Figure 7-4. Water level at the tide gauge at Inwood in the northeast corner of Jamaica Bay during Hurricane Sandy. Water levels remained above the major flood stage for several hours during this event. Data from the U.S. Geological Survey, National Water Information System (http://waterdata.usgs.gov/nwis/). Courtesy of Bernice Rosenzweig of CUNY Advanced Science Research Center.

academic institutions, and private companies. The meteorological stations at John F. Kennedy International Airport and Floyd Bennett Field are part of the National Oceanic and Atmospheric Administration's Historical Climatology Network, wherein instruments collect continuous data on basic meteorological variables such as surface temperature, precipitation, wind speed, and solar radiation, among many others. Data collected at these sites are subject to a common suite of quality assurance reviews and integrated into a database of daily data. The City College of New York manages the NYC MetNet Database (http://nycmetnet.ccny.cuny.edu) of surface observation sites managed by public and private agencies in the metropolitan region, with several sites located in the Jamaica Bay watershed. However, further study will be necessary to harmonize and adapt the data from the various MetNet sites to support climate change–related monitoring. Researchers can deploy meteorological stations, such as the one recently placed in the center of Jamaica Bay (figure 7-5), and collect data about storm events (figure 7-6).

Remote sensing data, including the quantitative precipitation estimates from the WSR-88D dual polarization radar at Upton, New York, and thermal imagery from the LandSat satellite, provide important information for the analysis of precipitation (Cunha et al., 2013) and urban heat island distributions (Rosenzweig et al. 2005) that may help fill gaps between meteorological stations. However, observational data from continuous monitoring sites are still needed for the calibration and validation of these data sets (Coll et al., 2010; Smith et al., 2007; Einfalt et al., 2004; Schott et al., 2001).

Figure 7-5. A meteorological mast deployed on the north side of Black Bank near the National Park Service visitor contact station at the Jamaica Bay Wildlife Refuge. Representative measurements are shown in figure 7-6. Courtesy of Christopher J. Zappa of Lamont-Doherty Earth Observatory, Columbia University.

Water and Sediment Quality Indicators

Because of the Clean Water Act and other regulatory considerations, plus the health and desirability of living near Jamaica Bay, issues of water quality and sediment supply and distribution are also critical to building resilience.

Policy Relevance and Analytical Soundness

The water quality of Jamaica Bay is influenced by a variety of factors, including discharges from wastewater treatment plants, runoff from impervious surfaces, combined sewer overflows (CSO), and historical change in geomorphology, as well as meteorological and tidal conditions, as discussed in chapters 4 and 5. With the exclusion of the CSO tributaries, Jamaica Bay is classified as a Class SB saline water body, with ambient concentrations of fecal coliform, enterococci, chlorophyll a, and dissolved oxygen that are suitable for

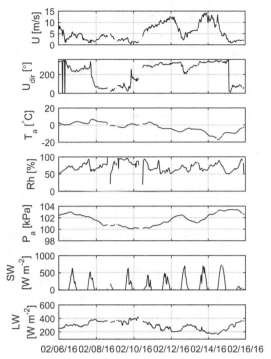

Figure 7-6. Weather measurements made at Jamaica Bay, over ten days in February 2016, during passage of winter storm Olympia. The plots show, from top to bottom, wind speed, wind direction, temperature, relative humidity, air pressure, and short-wave and long-wave radiation. Collectively, these data tell the story of the storm. A "quiet" period precedes the storm, followed by rapidly shifting wind directions, large wind speed variability, and sustained wind speeds during the storm itself. A sharp decrease in air temperature is associated with the low pressure system passing and the atmosphere being replenished with cold clear air. Courtesy of Christopher J. Zappa of Lamont-Doherty Earth Observatory, Columbia University.

fish propagation and primary human contact such as swimming. However, water quality conditions in Jamaica Bay can vary with location and time. Several hypoxia events have been recorded in recent years, particularly in the northeastern sections of the bay (Wallace et al., 2014). Concentrations of chlorophyll a, an indicator of phytoplankton concentrations and eutrophic conditions, can also be highly variable. Baseline water quality in some of the tributaries receiving CSO inputs is lower, and these water bodies are classified as Class I by the New York State Department of Environmental Conservation—suitable for secondary human contact such as boating, but not for swimming (New York City Department of Environmental Protection (NYCDEP), 2011).

As a result of the high population densities of its watershed, Jamaica Bay receives very high loads of nitrogen, which are primarily conveyed to the bay through the wastewater treatment plants, with additional contributions from CSO events, pumped groundwater for the dewatering of subway tunnels, landfill leachate, and groundwater flow (Benotti et al., 2007). This high loading of nitrogen is a major contributor to eutrophication of the bay, with subsequent algal blooms, reduced dissolved oxygen levels, and other ecological consequences (NYCDEP, 2007; Benotti et al., 2007). Several studies (including Deegan et al., 2012; Turner et al., 2009; and Benotti et al., 2007) have also implicated the high nitrogen input to the bay in the acceleration of salt marsh loss, though other possible mechanisms for marsh loss have also been identified (see discussion in chapter 5). As a result of all these factors, water quality needs to be monitored in conjunction with efforts to reduce nitrogen loading (NYCDEP, 2007, 2011).

Pharmaceuticals and personal care products have been recognized as "emerging contaminants;" these may be particularly important in Jamaica Bay due to the very high population densities in its watershed. In a recent study (Benotti and Brownawell, 2007), pharmaceuticals and selected major human metabolites were found to be ubiquitous in Jamaica Bay, though at very low concentrations. Future monitoring is necessary to understand how human activities and engineering practices may influence these concentrations and to determine if the low ambient concentrations of pharmaceuticals and personal care products have any impact on the biodiversity and ecological systems of the bay.

Sediment loading to Jamaica Bay has been greatly altered by human activities over the twentieth century. Jamaica Bay used to receive sediment from tidal creeks that fringe its perimeter; now it likely receives very little inasmuch as most of these creeks have since been filled (NYCDEP, 2007) and much of the watershed is now covered with an impervious layer. The westward extension and stabilization of the Rockaway Peninsula in the early part of the twentieth century may have resulted in reduced transport of offshore sediments into the bay (Hartig et al., 2002; Gordon et al., 2000). The input of sediments to Jamaica Bay may be an important resilience indicator that should be monitored by recording suspended material within the water column.

Circulation will be influenced by basin engineering, which can even overwhelm the impact of climate change on estuarine processes (Chant, 2002; Chant et al., 2011). For example, the deepening of the navigational channels by dredging has been shown in many systems to influence the effects of sea level rise on tidal processes and salt water intrusion (Talke and de Swart, 2006). There have been extensive changes to the shoreline of Jamaica Bay due to landfilling and urban development throughout the twentieth century (NYC DEP, 2007; see chapter 4, figure 4-1). Previous studies in Jamaica Bay have shown that tidal ranges in the bay have been amplified largely as a result of engineering changes during the twentieth century (Swanson and Wilson, 2008). To understand the combined impacts of climate change, dredging, and other engineering practices, indicators such as mean low-level water, tidal amplitude, and retention times at key locations throughout the bay should be monitored. Such studies can allow one to examine interactions between temperature and salinity that are important for biological use and functions (figure 7-7).

Sea level rise is expected to cause further changes to the shoreline as low-lying areas become inundated (Gornitz et al., 2001). Thus, changes in the geomorphology

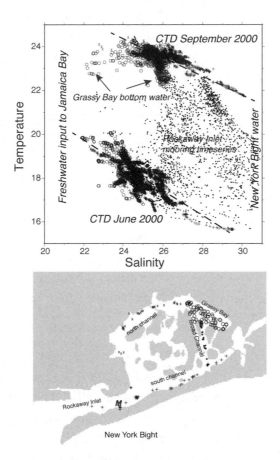

Figure 7-7. The temperature to salinity (T/S) relationship within Jamaica Bay observed from June through September 2000. The symbols on the upper panel correspond to locations shown on the lower panel. Data were collected by two boat surveys in June and September and continuously by a mooring within Rockaway Inlet (site M on map). The T/S plot shows the blend of the more saline, cooler coastal water from the New York Bight entering Jamaica Bay and mixing fresher, warmer waters over the course of the summer. The vigor of the exchange at Rockaway Inlet suggests residence time of water within Jamaica Bay of approximately seven days. However, within Grassy Bay, the deepest part of Jamaica Bay because of dredging for the airport, the waters near the bottom retain cooler waters the longest. CTD refers to the instrument used to measure conductivity, temperature and pressure in June and September 2000. Courtesy of Arnold Gordon of Lamont-Doherty Earth Observatory, Columbia University.

and topography of the shoreline will be important indicators for understanding the resilience of Jamaica Bay, and as such require monitoring through remote sensing and field surveys.

An additional factor stressing the marshes in Jamaica Bay is eutrophication (see discussion above and in chapters 4 and 5). In response to excess nutrients, marsh grasses tend to reduce their root biomass while increasing aboveground biomass, thus making the plant structure unstable and susceptible to collapse. The loss of the stabilizing plant biomass on the salt marsh increases the vulnerability of the marsh to erosive processes leading to marsh loss (Deegan et al., 2012). Nutrient levels in Jamaica Bay are controlled by the loadings of nutrients to the bay, biological processes, the geomorphological configuration of the bay shores and bathymetry, and flushing with the coastal ocean. Therefore, like models of salt marsh sustainability, models of water quality of Jamaica Bay require a high-quality data set of the critical physical and biogeochemical quantities that control water quality (see discussion in chapter 8).

The significance of interactions between Jamaica Bay waters and upland groundwater, and particularly how these interactions may be influenced by sea level rise, are less well understood (although see Misut and Voss, 2007). Preliminary studies of barrier island groundwater in Maryland indicate that even modest increases in sea level rise could significantly affect the water tables of barrier islands (Masterson et al., 2014). In the upland watersheds of Jamaica Bay, rising sea levels may also result in heightened water tables, with implications for subterranean infrastructure and runoff during intense rain events (NYS Sea Level Rise Task Force, 2010).

Measurability and Monitoring

Several government agencies monitor surface water quality in Jamaica Bay, including the NYCDEP, the Town of Hempstead Department of Environmental Conservation and Waterways, the New York State Department of Environmental Conservation, the New York–New Jersey Harbor Estuary Program, the Interstate Environment Commission, and the U.S. Geological Survey. However, to better understand water quality processes in Jamaica Bay, there is an urgent need for continuous, in situ monitoring of water quality in Jamaica Bay, as mentioned previously. This monitoring should include the biochemical observations of the ambient water column, as well as measurements of loading of pollutants delivered to the bay through CSOs from stormwater runoff.

Land Use and Land Cover Indicators

Land use and land cover change are probably the most important ways in which human beings affect Jamaica Bay (chapter 5, table 5-1). Today people use the land primarily for housing and for work, the very definition of urbanization. These factors are constantly changing within the Jamaica Bay watershed, with consequences for social and ecological resilience of the area.

Policy Relevance and Analytical Soundness

Like the historic case of the Jamaica Bay oyster, there is concern that the observed loss of salt marsh in Jamaica Bay may be a signal of a contemporary regime shift (see chapter 5). Most of the tidal wetlands of Jamaica Bay have been lost due to human activities, most notably through land filling (color plates I and V). In 1900, marshes covered 16,183 acres (6,549 ha) of the total area of Jamaica Bay, which included marsh islands as well as extensive shoreline marshes that extended beyond the Belt Parkway. By 1970 only 4,000 acres (1,619 ha) of marshland remained, with nearly all of the shoreline marsh lost to landfilling and urban development (Gordon et al., 2000; Englebright, 1975). Currently, the areal extent of the remaining marsh islands in Jamaica Bay's interior is decreasing at an accelerating rate, with important implications for biodiversity, coastal flood protection, and recreation in the bay. The areal extent and geomorphology (elevation, number of fractures, geometry of creeks, etc.) of the remaining salt marshes are, therefore, critical resilience indicators (chapter 5, figure 5-7).

Models of marsh dynamics pit the supply of sediment against sea level rise in determining the sustainability of salt marshes (Hartig et al., 2002; Fagherazzi et al., 2013; Mariotti and Fagherazzi, 2013). Although sea level rise is largely an external variable, the supply/loss of sediment to the marshes is controlled by wave action that can erode the marsh edge, by tidal and estuarine circulation that transports sediments throughout the system and coastal ocean (Ralston et al., 2013), and by internal marsh dynamics that influence the production and accumulation of organic matter (D'Alpaos et al., 2007). Wave action and tidal and estuarine circulation have likely been altered by basin engineering and thus have modified the supply of sediment to the marshes. Together with predictions of increased storminess (e.g., "Superstorm" Sandy) and accelerated sea level rise, understanding the sediment dynamics of Jamaica Bay is essential in developing predictive models of marsh sustainability in this system (Fagherazzi et al., 2012).

The resilience of Jamaica Bay and its watershed is also highly dependent on the distribution, character, and operation of human infrastructure, including residential buildings, businesses, roads and bridges, wastewater treatment plants, CSOs, sea walls, and green infrastructure, such as stormwater bioretention and blue roofs (NYCDEP, 2007). In turn, the resilience of the Jamaica Bay system will also affect the operation and social benefits provided by this infrastructure. For example, sea level rise may result in flooding of low-lying and subterranean infrastructure if adaptation measures are not employed. Potential variables that may provide insight for understanding system resilience include the volume and chemistry of wastewater treatment plant effluent; the frequency, volume, and chemistry of CSOs; the use and performance of green infrastructure; the distribution and types of shore-hardening infrastructure, such as sea walls; pumping rates of subterranean infrastructure (such as subway tunnels and basements); and the frequency of inundation of low-lying infrastructure (NYC Special Initiative for Rebuilding and Resiliency [SIRR], 2013). At present, most of this information is currently available, but it is collected and managed by a diverse group of city, state, and federal government agencies (including the NYCDEP, the Metropolitan Transit Authority, and the U.S. Army Corps of Engineers), private-sector consulting firms, and property owners. New York City Department of Parks and Recreation and the National Park Service regularly monitor wetland extent (Hartig et al., 2002; Gateway National Recreation Area, 2007). Additional work will be required to gain access to this information, to ensure that it continues to be collected consistently over time, and to harmonize it for use in resilience indication.

Measurability and Monitoring

There are many potential approaches for measuring changes in ecosystem extent, including the use of orthorectified imagery and other remote sensing data, complemented with regular field surveys. A number of permanent plots have been established for monitoring vegetation over time around Jamaica Bay. Sediment accretion rate is one important potential indicator, and can be measured at multiple locations throughout Jamaica Bay using several methods. The feldspar marker horizon technique, described in Cahoon and Turner (1998), can be used to measure contemporary accretion rates from soil core samples. Radiometric techniques can also be used to estimate long-term historic accretion rates from soil cores (Roman et al., 1997). Important indicators related to salt marsh erosion include soil water content, percent organic matter, particle size, and microbial decomposition rate. These measurements can be obtained through the seasonal collection of soil cores from the remaining marsh islands, along with push-pull pumping tests (Addy et al., 2002).

The redox chemistry of the shallow groundwater of salt marshes also plays an important role in its nutrient cycling, biodiversity, geomorphology, and the fate and transport of pollutants (Howes et al., 1986; Valiela and Teal, 1979). Low groundwater redox potential is associated with decreased productivity of salt marsh cordgrass (*Spartina alteriflora*) due to the reduced uptake of nutrients, increased respiratory carbon loss, and toxicity of accumulating hydrogen sulfide (H_2S; Howes et al., 1986). Recent studies have implicated excessive sulfide concentrations in salt marsh dieback in other East Coast estuaries (Alber et al., 2008), however, further study is required to determine conclusively whether the altered redox chemistry was the cause or an effect of the marsh dieback (Alber et al., 2008). Redox zonation may also play a role in the fate and transport of metals in Jamaica Bay sediments (Beck et al., 2009), with implications for salt marsh plants and water quality (Bentley et al., 2006; Xu and Jaffe, 2006; DeLaune et al., 1983). Because the redox chemistry of Jamaica Bay marshes may be affected by changes in water levels due to sea level rise or by changes in water column temperature and chemistry (Spalding and Hester, 2007; Bertness et al., 2002), the redox profile of salt marsh pore waters may be an important resilience indicator for Jamaica Bay that, to date, has been poorly characterized. Groundwater redox chemistry can be measured through field monitoring studies that include the use of passive dialysis samplers (El Bishlawi et al., 2013; Koretsky et al., 2003). These samplers provide monthly data on the profile of terminal electron acceptors and other redox-associated constituents. These samplers can be deployed at multiple sites in Jamaica Bay and at reference sites in the region that are not experiencing salt marsh loss.

Biodiversity and Species Abundance Indicators

As discussed in detail in chapter 5, Jamaica Bay is a unique urban sanctuary for wildlife and flora. The National Park Service has a congressional mandate to manage the natural resources of the Jamaica Bay Wildlife Refuge, and the public in New York City and adjacent regions benefits from having access to a wide diversity of species that is managed on parkland by the Natural Resources Group of the New York City Department of Parks and Recreation. Much of what people value about the bay comes from the ways that wildlife value it as well.

Policy Relevance and Analytical Soundness

Biodiversity indicators should provide information on what species are present, their distribution, and their health status (Hofmann and Gaines, 2008). Previous studies have identified several indicator species for which particular focus may be warranted, including oysters (*Crassostrea virginica*; Zarnoch and Schreibman, 2012), ribbed mussels (*Geukensia demissa*; Franz, 2001), horseshoe crabs (*Limulus polyphemus*; Botton et al., 2006) and winter flounder (*Pseudopleuronectes americanus*; Augspurger et al., 1994).

Important plants of the Jamaica Bay marshes include cord grasses (*Spartina alterniflora* and *Spartina patens*), black grass (*Juncus gerardii*), salt grass (*Distichlis spicata*), marsh elder (*Iva frutescens),* and seaside goldenrod (*Solidago sempervirens*). The biomass and distributions of plants and of nutrients within the roots and shoots of marsh vegetation may be important resilience indicators that should be monitored. Important information may also be provided through characterization of the distributions of invasive or potentially degrading species such as *Phragmites australis* and sea lettuce (*Ulva lactuca;* Hartig et al., 2002).

Measurability and Monitoring

The National Park Service, as well as many local organizations and academic researchers, conducts field surveys in different parts of the region. However, to date, there have been few efforts made to synthesize the results of this work, but efforts to do so should be a priority. Remote sensing data sets, such as aerial photography, can provide an important source of fine-scale information on the ecosystems of the bay and how they are affected by climate change and other stressors (Morgan et al., 2010). Aerial photos for New York City are managed by the Department of Information Technology and Telecommunications. Other applications of remote sensing data include the use of hyperspectral and multispectral imagery from satellites and LIDAR elevation data sets to characterize high and low marsh vegetation and the use of remote sensing indices such as the normalized difference vegetation index to monitor changes in vegetation biomass and stress over time (Klemas, 2011a). Satellite and airborne measurements of spectral reflectance can also provide important information on spatial and temporal variability in algal blooms (Klemas, 2011b).

Genomics, the study of the entire genome of organisms, may also provide important new techniques to support resilience indicators. Genomic technologies can be used to detect the presence of previously unidentified microbes in environmental samples. The recent availability of this technology has provided an unprecedented view of microbial diversity in marine systems. Given the important role that microbes play in coastal ecosystems, the use of these techniques can shed significant light on the health of ecosystems and their changes through time and space due to environmental shocks and stresses (Hofmann and Gaines, 2008; DeLong and Karl, 2005).

The DNA microarray has also become an important new tool for understanding the response of organisms to multiple environmental stressors (Hofmann and Gaines, 2008). This technique provides a profile of gene expression—which genes are turned on and off—in environmental samples and can provide insight on how the environmental factors may affect the genome. Previous studies have revealed significant changes in the gene expression of aquatic organisms with changes in physical factors such as temperature (Buckley, 2007).

Community Resilience Indicators

Next and finally, but not last in importance, we discuss measures of community resilience.

Policy Relevance and Analytical Soundness

The social systems of Jamaica Bay residents are inextricably linked to Jamaica Bay's ecosystems and resilience (see chapters 3, 6, and 11). Many Jamaica Bay communities, such as the dockside neighborhoods of Broad Channel, Old Howard Beach, and Mill Basin, and the beach communities of Far Rockaway, are centered on the recreational services provided by the bay, and the bay plays an important role in the economies of these communities. Jamaica Bay also provides important cultural and spiritual services for diverse communities throughout its watershed (Kornblum and van Hooreweghe, 2011). Potential indicators that can be used to describe the use, status, and resilience of these services include those that measure direct human interaction with the bay, such as frequency of beach visits, boating, or other recreational activities. Economic indicators such as annual revenue from tourism or, alternatively, damages from severe events, should also be considered.

As discussed in chapter 5, the organization of Jamaica Bay's social systems also plays an important role in its resilience to disturbances (Adger, 2000). Demographic characteristics, including poverty, gender, and age, have been identified as key social indicators that can be used to determine a location's vulnerability to extreme events (Rygel et al., 2006). Other indicators commonly used to assess social vulnerability include those that identify special needs populations or those that lack the "social safety nets" that aid in resilience to extreme events, such as percentage of the population with disabilities, or that are non-English-speaking, homeless, or tourists that are unfamiliar with an area (Cutter et al., 2003). These indicators are often composited into a social vulnerability index that describes the overall vulnerability of a population to hazardous events due to social factors. For example, from 1960 to 2010, Kings County, New York (Brooklyn), was one of only three counties in the United States included as the top twenty-five most vulnerable in each decade (Cutter and Finch, 2008). In addition to assessing vulnerability to potential events, the impacts of extreme events and effectiveness of resilience efforts can be tracked through basic social metrics, such as the numbers of injuries and fatalities resulting from events that occur.

At the same time, social systems can play a key role in mitigating vulnerability to natural hazards and disturbances (see chapter 10). Community social networks can play an important role in building resilience (Tompkins and Adger, 2004). The importance of networks was demonstrated in the aftermath of Hurricane Sandy, when organized volunteers played a critical role in assisting the elderly and disabled who were stranded in their homes (NYC SIRR, 2013) and in cleanup and aid-distribution efforts in the aftermath of the storm (Williams, 2014). Although the role of these social networks, and potential quantitative indicators to represent them, are still a relatively recent area of research (Newman and Dale, 2005), their potential importance makes them worthy of consideration in the development of a Jamaica Bay resilience indicators system, and the social science community should identify potential metrics that meet the three criteria of appropriate resilience indicators. Potential candidates may include the rate of demographic changes within a neighborhood, number of local community organizations, and rate of membership of these organizations in New York City's Community Resilience Planning Committee.

Measurability and Monitoring
Potential sources of socioeconomic data include the U.S. Census, which provides demographic information at the census block level and economic information at the census

tract level. New York City and Company, the city's official marketing and tourism organization (www.nycandcompany.org/research), collects and analyzes data on travel and tourism in New York City, including Jamaica Bay and its watershed. The New York City Parks Department and the National Park Service also collect information on beach visits and other recreational activities in Jamaica Bay and Gateway National Recreational Area. Damages from severe events are estimated by a number of public entities, including the New York City Office of Management and Budget, the Federal Insurance Services Office—Property Claims Service, the Federal Emergency Management Agency, and the National Flood Insurance Program, along with private entities.

Conclusions: Indicators and Monitoring for Resilience

We have proposed an initial description of climate hazard, water and sediment quality, land use and land cover, biodiversity and species abundance, and community resilience indicators that we believe will provide an important foundation for a resilience analysis system for Jamaica Bay. Similar sets of indicators could be adapted for other urban coastal watersheds.

However, this chapter represents only the beginning of this process—through a dialogue among physical scientists, social scientists, and stakeholders, a comprehensive list of key contemporary ecosystem services provided by Jamaica must be developed. Once this list has been established the relationship between the provision of these services and measureable components of the Jamaica Bay system must be determined, which will serve as a subset of the resilience indicators for Jamaica Bay. Potential shocks and stresses to these components must also be identified, and measurable indicators describing their occurrence and magnitude must be determined.

Resilience analysis requires information drawn from an array of sustained observations that capture the spatial and temporal patterns of the Jamaica Bay system—the attributes that contribute to overall ecosystem metabolism. Such observations may consist of in situ sensors at key sites to assess the bay's response to external and internal forcing, as well as use of remotely gathered data—for example, satellite data. Focused process studies must also be embedded in the in situ observational array as needed to better understand the governing forces that shape the Jamaica Bay system and to identify potentially important disturbances. All of this needs to be closely coupled with modeling techniques, which are the discussion of the following chapter.

An extensive web of environmental monitoring systems currently collects data that can support climate indicators monitoring for Jamaica Bay. However, remote sensing data sets must be synthesized and site-based monitoring procedures must be harmonized to allow for the spatial and temporal comparison needed by resilience indicators. It will also be necessary to integrate the existing observations of physical climate change parameters with updated information on infrastructure and demographic indicators. As we will discuss below, some key gaps and limitations in existing monitoring systems must be filled by future environmental monitoring efforts by the Science and Resilience Institute at Jamaica Bay to support a comprehensive resilience analysis.

Process studies, supplemented by numerical modeling, may play an important role in identifying the relationships between system components, ecosystem services, and system stresses and shocks (chapter 8). Process-based studies are critical to support the development of a comprehensive portfolio of resilience indicators for Jamaica Bay and to support the interpretation of indicator monitoring data for resilience analysis. The most comprehensive process studies to date in Jamaica Bay were conducted through the integrated reconnaissance of the physical and biogeochemical characteristics of Jamaica Bay, a study conducted by the National Park Service and the Columbia University Earth Institute (Gordon et al., 2000). Through this work, a number of key physical and biogeochemical characteristics have been collected on Jamaica Bay (Gordon et al., 2000) that provided new insight on the complex system. This includes the significant transport of coarse sediment in the channels of Jamaica Bay, with Grassy Bay serving as a significant sink for sediment that might otherwise provide a source for accretion in Jamaica Bay salt marshes. This study also showed that flushing times vary in different regions of the bay, and estimates using two independent methods yielded a flushing time of approximately one week for the upper 16.4 feet (5 m) of Grassy Bay. This study also identified multiple sources of freshwater to the bay, including sewage treatment plant effluent surface run-off and the Hudson River plume. Nitrogenous nutrients remained abundant throughout the summer, and researchers noted periods of suboxic conditions at the sediment-water interface in Grassy Bay, which has been made deeper than other parts of Jamaica Bay by dredging. During hypereutrophic conditions, the phytoplankton appeared to be limited by the availability of carbon dioxide.

Although there are many potential sources of observational data that can be used in the development of a resilience monitoring system, additional work is necessary to ensure that we fill data gaps and to ensure that the data are consistent and comparable over long time scales, throughout Jamaica Bay and with other analogue sites. For some measures, such as water quality observations collected by NYCDEP and the National Park Service, we have relatively long-term records. For other kinds of observations, covering a wide range of data, we need to look toward sustained observational arrays. Scientists with the Long-Term Ecological Research Network of the National Science Foundation (http://www.lternet.edu) have developed frameworks to support sustained, integrated ecosystem research and make the data accessible to both scientific researchers and stakeholders. Such frameworks might provide a model for Jamaica Bay and other urban watersheds.

References

Addy, K., Kellogg, D.Q., Gold, A.J., Groffman, P.M., Ferendo, G., and Sawyer, C. 2002. In situ push–pull method to determine ground water denitrification in riparian zones. *Journal of Environmental Quality* 31(3): 1017–1024.

Adger, W.N. 2000. Social and ecological resilience: Are they related? *Progress in Human Geography* 24(3): 347–364.

Alber, M., Swenson, E.M., Adamowicz, S.C., and Mendelssohn, I.A. 2008. Salt marsh dieback: An overview of recent events in the US. *Estuarine, Coastal and Shelf Science* 80(1): 1–11.

Anthony, A., Atwood, J., August, P.V., Byron, C., Cobb, S., Foster, C., et al. 2009. Coastal lagoons and climate change: Ecological and social ramifications in the US Atlantic and Gulf Coast ecosystems. *Ecology and Society* 14(1): art. 8. Accessed at: http://www.ecologyandsociety.org/vol14/iss1/art8/.

Augspurger, T.P., Herman, R.L., Tanacredi, J.T., and Hatfield, J.S. 1994. Liver lesions in winter flounder (*Pseudopleuronectes americanus*) from Jamaica Bay, New York: Indications of environmental degradation. *Estuaries* 17(1): 172–180.

Beck, A.J., Cochran, J.K., and Sañudo-Wilhelmy, S.A. 2009. Temporal trends of dissolved trace metals in Jamaica Bay, NY: Importance of wastewater input and submarine groundwater discharge in an urban estuary. *Estuaries and Coasts* 32(3): 535–550.

Benotti, M.J., and Brownawell, B.J. 2007. Distributions of pharmaceuticals in an urban estuary during both dry- and wet-weather conditions. *Environmental Science & Technology* 41(16): 5795–5802.

Benotti, M.J., Abbene, I., and Terracciano, S.A. 2007. Nitrogen loading in Jamaica Bay, Long Island, New York: Predevelopment to 2005. U.S. Geological Survey.

Bentley, S., Thibodeaux, L., Adriaens, P., Li, M.Y., Romero-González, M., et al. 2006. Physicochemical and biological assessment and characterization of contaminated sediments. In: *Assessment and Remediation of Contaminated Sediments*. 83–136. Netherlands: Springer.

Bertness, M.D., Ewanchuk, P.J., and Silliman, B.R. 2002. Anthropogenic modification of New England salt marsh landscapes. *Proceedings of the National Academy of Sciences* 99(3): 1395–1398.

Black, F.R. 1981. Jamaica Bay: A History. Cultural Resource Management Study 3. Washington, D.C. Available from http://www.nps.gov/history/history/online_books/gate/jamaica_bay_hrs.pdf.

Blackford, E.G., 1884. Report of the oyster investigation and of survey of oyster territory. Report of the oyster investigation and shell-fish commission. State of New York, Albany.

Botton, M.L., Loveland, R.E., Tanacredi, J.T., and Itow, T. 2006. Horseshoe crabs (*Limulus polyphemus*) in an urban estuary (Jamaica Bay, New York) and the potential for ecological restoration. *Estuaries and Coasts* 29(5): 820–830.

Buckley, B.A. 2007. Comparative environmental genomics in non-model species: Using heterologous hybridization to DNA-based microarrays. *Journal of Experimental Biology* 210(9): 1602–1606.

Cahoon, D.R., and Turner, R.E. 1989. Accretion and canal impacts in a rapidly subsiding wetland II. Feldspar marker horizon technique. *Estuaries* 12(4): 260–268.

Chant, R.J. 2002. Secondary circulation in a region of flow curvature: Relationship with tidal forcing and river discharge. *Journal of Geophysical Research: Oceans* 107(C9):

14-1–14-11.

Chant, R.J., Fugate, D., and Garvey, E. 2011. The shaping of an estuarine Superfund site: Roles of evolving dynamics and geomorphology. *Estuaries and Coasts* 34: 90–105, doi:10.1007/s12237-010-9324-z.

City of New York. 2015. *OneNYC: The Plan for a Strong and Just City*. Accessed at: http://www.nyc.gov/html/onenyc/downloads/pdf/publications/OneNYC.pdf.

Coll, C., Galve, J.M., Sanchez, J.M., and Caselles, V. 2010. Validation of Landsat-7/ETM+ thermal-band calibration and atmospheric correction with ground-based measurements. *IEEE Transactions on Geoscience and Remote Sensing,* 48(1): 547–555.

Cumming, G.S., Barnes, G., Perz, S., Schmink, M., Sieving, K.E., et al. 2005. An exploratory framework for the empirical measurement of resilience. *Ecosystems* 8(8): 975–987.

Cutter, S.L., Boruff, B.J., and Shirley, W.L. 2003. Social vulnerability to environmental hazards. *Social Science Quarterly* 84(2): 242–261.

Cutter, S.L., and Finch, C. 2008. Temporal and spatial changes in social vulnerability to natural hazards. *Proceedings of the National Academy of Sciences* 105(7): 2301–2306.

D'Alpaos, A., Lanzoni, S., Marani, M., and Rinaldo, A. 2007. Landscape evolution in tidal embayments: Modeling the interplay of erosion, sedimentation, and vegetation dynamics. *Journal of Geophysical Research: Earth Surface (2003–2012)* 112(F1).

Deegan, L.A., Johnson, D.S., Warren, R.S., Peterson, B.J., Fleeger, J.W., et al. 2012. Coastal eutrophication as a driver of salt marsh loss. *Nature* 490(7420): 388–392.

DeLaune, R.D., Smith, C.J., and Patrick, W. 1983. Relationship of marsh elevation, redox potential, and sulfide to *Spartina alterniflora* productivity. *Soil Science Society of America Journal* 47(5): 930–935.

DeLong, E.F., and Karl, D.M. 2005. Genomic perspectives in microbial oceanography. *Nature* 437(7057): 336–342.

Einfalt, T., Arnbjerg-Nielsen, K., Golz, C., Jensen, N.E., Quirmbach, M., et al. 2004. Towards a roadmap for use of radar rainfall data in urban drainage. *Journal of Hydrology* 299(3): 186–202.

El Bishlawi, H., Shin, J.Y., and Jaffe, P.R. 2013. Trace metal dynamics in the sediments of a constructed and natural urban tidal marsh: The role of iron, sulfide, and organic complexation. *Ecological Engineering* 58: 133–141.

Englebright, S. 1975. *Jamaica Bay: A Case Study of Geo-Environmental Stresses: A Guidebook to Field Excursions.* Hempstead, NY: New York State Geological Association, Hofstra University.

Fagherazzi, S., Kirwan, M.L., Mudd, S.M., Guntenspergen, G.R., Temmerman, S., et al. 2012. Numerical models of salt marsh evolution: Ecological, geomorphic, and climatic factors. *Reviews of Geophysics* 50(1). doi:10.1029/2011RG000359.

Fagherazzi, S., Mariotti, G., Wiberg, P.L., and McGlathery, K.J. 2013. Marsh collapse does not require sea level rise. *Oceanography* 26(3): 70–77. http://dx.doi.org/10.5670/oceanog.2013.47.

Franz, D.R., 1982. An historical perspective on mollusks in lower New York Harbor, with emphasis on oysters. In: *Ecological Stress and the New York Bight: Science and Management*, Meyer, G.F. (ed.). 181–197. Columbia, SC: Estuarine Research Federation.

Franz, D.R. 2001. Recruitment, survivorship, and age structure of a New York ribbed mussel population (*Geukensia demissa*) in relation to shore level—a nine-year study. *Estuaries* 24(3): 319–327.

Gateway National Recreation Area, 2007. An Update on the Disappearing Salt Marshes of Jamaica Bay, New York. Staten Island, NY: National Park Service, US Department of the Interior, Gateway National Recreation Area.

Gemmrich, J.R., and Farmer, D.M. 2004. Near-surface turbulence in the presence of breaking waves. *Journal of Physical Oceanography* 34: 1067–1086.

Gordon, A.L., Bell, R., Carbotte, S., Flood, R., Hartig, E., et al. 2000. Integrated Reconnaissance of the Physical and Biogeochemical Characteristics of Jamaica Bay. Accessed at: http://www.ldeo.columbia.edu/res/div/ocp/projects/jamaicabay.shtml.

Gornitz, V., Couch, S., and Hartig, E.K. 2001. Impacts of sea level rise in the New York City metropolitan area. *Global and Planetary Change* 32(1): 61–88.

Grambo, G., and Vega, C., 1984. Jamaica Bay: a site study. U.S. Geological Survey, Long Island Division, Coram, NY.

Hartig, E.K., Gornitz, V., Kolker, A., Mushacke, F., and Fallon, D. 2002. Anthropogenic and climate-change impacts on salt marshes of Jamaica Bay, New York City. *Wetlands* 22(1): 71–89.

Hendrick, D.M., 2006. *Jamaica Bay*. Charleston, SC: Arcadia Publishing.

Hofmann, G.E., and Gaines, S.D. 2008. New tools to meet new challenges: Emerging technologies for managing marine ecosystems for resilience. *BioScience* 58(1): 43–52.

Howes, B.L., Dacey, J.W.H., and Goehringer, D.D. 1986. Factors controlling the growth form of *Spartina alterniflora:* Feedbacks between above-ground production, sediment oxidation, nitrogen and salinity. *Journal of Ecology,* 74(3): 881–898.

Interstate Sanitation Commission. 1937. Annual report of the Interstate Sanitation Commission for the year 1937. Trenton, NJ: Interstate Sanitation Commission.

Jacob, K., Blake, R., Horton, R., Bader, D., and O'Grady, M. 2010. Climate Change Adaptation in New York City: Building a Risk Management Response: New York City Panel on Climate Change 2010 Report, Chapter 7: Indicators and monitoring. *Annals of the New York Academy of Sciences* 1196: 127–142. doi:10.1111/j.1749-6632.2009.05321.x.

Klemas, V. 2011a. Remote sensing of wetlands: Case studies comparing practical techniques. *Journal of Coastal Research* 27(3): 418–427.

Klemas, V. 2011b. Remote sensing of algal blooms: An overview with case studies. *Journal of Coastal Research* 28(1A): 34–43.

Koretsky, C.M., Moore, C.M., Lowe, K.L., Meile, C., DiChristina, T.J., and Van Cappellen, P. 2003. Seasonal oscillation of microbial iron and sulfate reduction in saltmarsh sediments (Sapelo Island, GA, USA). *Biogeochemistry* 64(2): 179–203.

Kornblum, W., and van Hooreweghe, K. 2011. Jamaica Bay Ethnographic Overview and Assessment. Northeast Region Ethnography Program. National Park Service, Boston, MA.

Levinton, J., Doall, M., and Allam, B. 2013. Growth and mortality patterns of the eastern oyster *Crassostrea virginica* in impacted waters in coastal waters in New York, USA. *Journal of Shellfish Research* 32: 417–427.

MacKenzie, C.L. Jr. 1996. History of Oystering in the United States and Canada, Featuring the Eight Greatest Oyster Esutaries. *Marine Fisheries Review* 58: 1–78.

McLaughlin, J. 2014. Expansion of oyster pilot study within Jamaica Bay. Presentation to the Jamaica Bay Task Force, New York, NY. Accessed at: http://www.slideshare.net/ecowatchers/jbtf-oyster-presentation.

MacCready, P., and Geyer, W.R. 2010. Advances in estuarine physics. *Annual Review of*

Marine Science 2: 35–58.

Mariotti, G., and Fagherazzi, S. 2013. A two-point dynamic model for the coupled evolution of channels and tidal flats. *Journal of Geophysical Research: Earth Surface* 118(3): 1387–1399.

Masterson, J.P., Fienen, M.N., Thieler, E.R., Gesch, D.B., Gutierrez, B.T., and Plant, N.G. 2014. Effects of sea-level rise on barrier island groundwater system dynamics—ecohydrological implications. *Ecohydrology* 7(3): 1064–1071.

Misut, P.E., and Voss, C.I., 2007. Freshwater–saltwater transition zone movement during aquifer storage and recovery cycles in Brooklyn and Queens, New York City, USA. *Journal of Hydrology* 337: 87–103. doi:10.1016/j.jhydrol.2007.01.035.

Morgan, J.L., Gergel, S.E., and Coops, N.C. 2010. Aerial photography: A rapidly evolving tool for ecological management. *BioScience* 60(1): 47–59.

New York Times. 1921. Jamaica Bay, Foul With Sewage, Closed To Oyster Beds; 300,000 Bushels Gone. *New York Times*, New York.

Newman, L., and Dale, A. 2005. Network structure, diversity, and proactive resilience building: A response to Tompkins and Adger. *Ecology and Society* 10(1): r2.

New York City Panel on Climate Change (NPCC). 2013. Climate Risk Information 2013: Observations, Climate Change Projections and Maps. A report by the New York City Panel on Climate Change. June 2013. Accessed at: http://www.nyc.gov/html/planyc2030/downloads/pdf/npcc_climate_risk_information_2013_report.pdf.

New York City Department of Environmental Protection (NYC DEP). 2007. Watershed Protection Plan. New York City Department of Environmental Protection. Accessed at: http://nyc.gov/html/dep/html/dep_projects/jamaica_bay.shtml.

NYCDEP. 2011. Jamaica Bay Waterbody/Watershed Facility Plan. The New York City Department of Environmental Protection. October 2011. Accessed at: http://www.hydroqual.com/projects/ltcp/wbws/jamaica_bay/jamaica_bay_cover.pdf.

NYC SIRR. 2013. *A Stronger, More Resilient New York*. Report by the New York City Special Initiative for Rebuilding and Resiliency. Accessed at: http://www.nyc.gov/html/sirr/html/report/report.shtml.

New York State Sea Level Rise Task Force. 2010. Report to the Legislature. December 31, 2010. Accessed at: http://www.dec.ny.gov/docs/administration_pdf/slrtffinalrep.pdf.

Orton, P.M., McGillis, W.R., and Zappa, C.J. 2010. Sea breeze forcing of estuary turbulence and air-water CO_2 exchange. *Geophysical Research Letters* 37(L13603). doi:10.1029/2010GL043159.

Ramesh, R., Chen, Z., Cummins, V., Day, J., D'Elia, C., et al. 2016. Land-ocean interactions in the coastal zone: Past, present and future. *Anthropocene*. http://dx.doi.org/10.1016/j.ancene.2016.01.005.

Roman, C.T., Peck, J.A., Allen, J.R., King, J.W., and Appleby, P.G. 1997. Accretion of a New England (USA) salt marsh in response to inlet migration, storms, and sea-level rise. *Estuarine, Coastal and Shelf Science* 45(6): 717–727.

Rosenzweig, C., Solecki, W.D., Parshall, L., Chopping, M., Pope, G., and Goldberg, R. 2005. Characterizing the urban heat island in current and future climates in New Jersey. *Global Environmental Change Part B: Environmental Hazards* 6: 51–62. doi:10.1016/j.hazards.2004.12.001.

Rygel, L., O'Sullivan, D., and Yarnal, B. 2006. A method for constructing a social vulnerability index: An application to hurricane storm surges in a developed country. *Mitigation and Adaptation Strategies for Global Change* 11(3): 741–764.

Schott, J.R., Barsi, J.A., Nordgren, B.L., Raqueno, N.G., and De Alwis, D. 2001. Calibration of Landsat thermal data and application to water resource studies. *Remote Sensing of Environment* 78(1): 108–117.

Sigler, M., and Leibovitz, L. 1982. Acute toxicity of oil and bilge cleaners to larval American oysters (*Crassostrea virginica*). *Bulletin of Environmental Contamination and Toxicology* 29: 137–145.

Smith, J.A., Baeck, M.L., Meierdiercks, K.L., Miller, A.J., and Krajewski, W.F. 2007. Radar rainfall estimation for flash flood forecasting in small urban watersheds. *Advances in Water Resources* 30(10): 2087–2097.

Solecki, W., Rosenzweig, C., Blake, R., Sherbinin, A., Matte, T., et al. 2015. New York City Panel on Climate Change 2015 Report, Chapter 6: Indicators and Monitoring. *Annals of the New York Academy of Sciences* 1336(1): 89–106.

Spalding, E.A., and Hester, M.W. 2007. Interactive effects of hydrology and salinity on oligohaline plant species productivity: Implications of relative sea-level rise. *Estuaries and Coasts* 30(2): 214–225.

Swanson, R.L., and Wilson, R.E. 2008. Increased tidal ranges coinciding with Jamaica Bay development contribute to marsh flooding. *Journal of Coastal Research* 24(6): 1565–1569.

Talke, S.A., and de Swart, H.E. 2006. Hydrodynamics and morphology in the Ems/Dollard estuary: Review of models, measurements, scientific literature, and the effects of changing conditions. Institute for Marine and Atmospheric Research Utrecht (IMAU).

Tanacredi, J.T. 1977. Petroleum hydrocarbons from effluents: Detection in marine environment. *Journal of the Water Pollution Control Federation* 49: 216–226.

Tompkins, E.L., and Adger, W. 2004. Does adaptive management of natural resources enhance resilience to climate change? *Ecology and Society* 9(2): 10.

Turner, R.E., Howes, B.L., Teal, J.M., Milan, C.S., Swenson, E.M., and Goehringer-Toner, D.D. 2009. Salt marshes and eutrophication: An unsustainable outcome. *Limnology and Oceanography* 54(5): 1634.

Valiela, I., and Teal, J.M. 1979. The nitrogen budget of a salt marsh ecosystem. *Nature* 280(5724): 652–656.

Waldman, J., 2008. Research Opportunities in the Natural and Social Sciences at the Jamaica Bay Unit of Gateway National Recreation Area. Jamaica Bay Institute, New York.

Wallace, R.B., Baumann, H., Grear, J.S., Aller, R.C., and Gobler, C.J. 2014. Coastal ocean acidification: The other eutrophication problem. *Estuarine, Coastal and Shelf Science* 148: 1–13. doi:10.1016/j.ecss.2014.05.027.

Walsh, J., Wuebbles, D., Hayhoe, K., Kossin, J., Kunkel, K., et al. 2014. Chapter 2: Our Changing Climate. *Climate Change Impacts in the United States: The Third National Climate Assessment*, Melillo, J.M., Richmond, T.C., and Yohe, G.W. (eds). 19–67. U.S. Global Change Research Program. doi:10.7930/J0KW5CXT.

Williams, E. 2014. Social Resiliency and Superstorm Sandy. A report by the Association for Neighborhood and Housing Development. Accessed at: http://www.anhd.org/wp -content/uploads/2011/07/Social-Resiliency-and-Superstorm-Sandy-11-14.pdf.

Xu, S., and Jaffé, P.R. 2006. Effects of plants on the removal of hexavalent chromium in wetland sediments. *Journal of Environmental Quality* 35(1): 334–341.

Zappa, C.J., McGillis, W.R., Raymond, P.A., Edson, J.B., Hintsa, E.J., et al. 2007. Environmental turbulent mixing controls on the air-water gas exchange in marine and aquatic systems. *Geophysical Research Letters* 34(10): L10601. doi:10.1029/2006GL028790.

Zappa, C.J., Raymond, P.A., Terray, E., and McGillis, W.R. 2003. Variation in surface turbulence and the gas transfer velocity over a tidal cycle in a macro-tidal estuary. *Estuaries* 26(6): 1401–1415.

Zarnoch, C.B., and Schreibman, M.P. 2012. Growth and reproduction of eastern oysters, *Crassostrea virginica*, in a New York City estuary: Implications for restoration. Urban Habitats, 7(1). Accessed at: http://www.urbanhabitats.org/v07n01/easternoysters_full .html.

8

Computational Modeling of the Jamaica Bay System

Eric W. Sanderson, Philip Orton, Jordan R. Fischbach,
Debra Knopman, Hugh Roberts, William D. Solecki,
and Robert Wilson

Computational models are essential tools to support resilience planning for Jamaica Bay, or indeed anywhere (Hawes and Reed, 2006; Pickett et al., 2004; Walker et al., 2002; Gallopin, 2002). Models are simplifications of reality, constructed to highlight the interactions among physical, ecological, and social components of a system. Models connect observations with hypotheses and theories about how physical and social systems work, allowing scientists to articulate and test system understanding against data. Although there are physical and conceptual models, in the early twenty-first century, most models are deployed on computers and are increasingly used in distributed computing environments accessible through the Internet.

Computer models take inputs of digital data describing system conditions, transform them through logical expressions of relationships among inputs, and produce a set of digital outputs. In a resilience context, inputs might include landscape or climate descriptions, model syllogisms might describe disturbance processes that trigger changes, and outputs describe valued system attributes. For a coastal estuarine system such as Jamaica Bay, initial conditions might include bathymetry, topography, climate, and land use; disturbances might include storm events or changes in government policy; and outputs might describe performance metrics related to resilience goals, such as minimizing flood damage, enhancing biodiversity, and avoiding economic loss.

The flexible yet systematic structure of models makes them excellent platforms for synthesis and collaboration across disciplines. Models can be linked, where the outputs from one model become inputs to the next. For example, one model might simulate the frequency and strength of hurricane winds and waves, another might estimate the height and duration of flooding events caused by storm surge, and yet another might estimate the damaging effects of salt water flooding on buildings (e.g., Georgas et al., 2014).

Models require validation to provide confidence in the outputs. Validations are checks of the model results against independent data, collected either through direct observation or other models. The validation process can expose weaknesses, which may be related to uncertainty or incompleteness of inputs, or to fundamental lack of understanding. When models underperform, the discrepancies may suggest areas where further research and data collection would be most useful.

Computational models are particularly useful to support scenario planning (Alcamo, 2008; Schoemaker, 1995). Scenarios represent hypothesized alternative states of a system. Decisions about how to modify the landscape or enhance resilience often depend on exploratory analysis. Because uncertainty about the future is often large ("deep uncertainty"), scenarios enable resilience planning to proceed (Groves and Lempert, 2007).

Identifying scenarios relevant for near-term decisions can be difficult. Simulation models used in a robust decision-making framework have been helpful in other urban estuaries and might be relevant to Jamaica Bay (e.g., Fischbach et al., 2015; Groves et al., 2014). Scenarios might be constructed through variation of plausible realizations of uncertainty in the inputs—for example, different storm frequencies or rates of population change. Such model runs can be thought of as testing the effects of these exogenous drivers of the system. The effect of policy interventions such as the construction of a seawall or restoration of salt marsh can also be tested under various scenarios.

An important area of focus for resilience planning is how scenarios are generated: who gets to say which alternatives should be modeled? Traditionally, alternatives have been generated by experts or those with management authority, but increasingly there are calls (e.g., Adger, 2003; Reed, 2008) to democratize the process—for example, by having the public generate scenarios or interventions or even giving stakeholders access to models.

In Jamaica Bay, computational models are already used in decision-making contexts. Particular emphasis has been placed on models of oceanographic processes and water quality, as discussed below. Recent work by the RAND Corporation and partners has attempted to connect these models into an integrated modeling framework, based on their experience after Hurricanes Katrina and Rita (Groves et al., 2014). In a similar vein, recent work by the Wildlife Conservation Society's Visionmaker project for Jamaica Bay, described below, attempts to link computational models of the Jamaica Bay social-ecological system with efforts to democratize the planning process through an Internet-based scenario planning tool (http://Visionmaker.nyc).

In this chapter we review the use of computational models in resilience planning generally and particularly in the Jamaica Bay context (table 8-1). We begin with a discussion

Table 8-1. Selected computational models applicable to Jamaica Bay.

Model	Domain*	Dimensionality	Grid or Element Resolution	Outputs	Reference
ADCIRC/SWAN	FEMA Region 2	2-D x time	~70 m	flooding, waves	FEMA (2014); Orton et al. (2015b)
ADCIRC/SWAN	New York City	2-D x time	~40 m	flooding, waves	City of New York (2013)
Best Practices Model (BPM)	New York City metro region	0-D x time	road network	vehicular traffic	NYMTC (2016)
Coastal Louisi-ana Risk Assess-ment (CLARA)	Variable	2-D	tax parcel	flood hazard, flood loss	Fischbach et al. (2012)
Coupled Model Inter-comparison Project Phase 5 (CMIP5)	Earth	3-D x time	0.5–4 degrees	air temperature, precipitation, sea level	Horton et al. (2015)
Delft 3-D	Jamaica Bay	3-D x time	10 m and 50 m	circulation, flood-ing, sedimentation	Wang, H. (U.S. Geo-logical Survey, pers. comm.)
Department of City Planning Baseline Demo-graphic Model	New York City	0-D x time	boroughs	human population	Salvo et al. (2006)
Finite-Volume Coastal Ocean Model (FVCOM)	Jamaica Bay	3-D x time	~5–100 m	circulation, flood-ing, sedimentation	Wilson (2008)
Flood Damage Reduction Anal-ysis (HEC-FDA)	Jamaica Bay	2-D	variable	flood hazard, flood loss	Knopman and Fischbach (pers. comm.)
Hazus-Multi-Hazards (Hazus-MH)	Jamaica Bay	2-D	30 m	flood hazard, flood loss	FEMA (2014)
InfoWorks	New York City	2-D x time	2.4 m	stormwater flows	NYCDEP (2012)
Jamaica Bay Eutrophication Model (JEM)	Jamaica Bay 2012 (likely 2016 update)	3-D x time	20–200 m	water quality; oyster larvae transport	Hydroqual (2002, 2012)
Marsh Equi-librium Model (MEM)	Variable	0-D x time	point	marsh elevation and productiv-ity; carbon sequestration	Morris (2015)

Table 8-1. continued

Model	Domain*	Dimensionality	Grid or Element Resolution	Outputs	Reference
Saturated-Unsaturated Transport (Sutra)	western Long Island	3-D x time	~200–1,000 m	groundwater flows	Misut and Voss (2007)
Sea Level Affecting Marshes Model (SLAMM)	Variable	3-D x time	variable	marsh distribution	Larsen, M. (pers. comm.)
Stevens Inst. Estuarine and Coastal Ocean hydrodynamic Model (sECOM)	Jamaica Bay flooding	2-D x time	30 m	circulation, tides, storm surge, flooding	Orton et al. (2015a)
sECOM	Jamaica Bay using JEM2016 grid	3-D x time	20–200 m	circulation, tides, residence time	Marsooli et al. 2016b
Visionmaker	New York City	2-D	10 m	carbon flows (including carbon emissions, sequestration, and organic waste); water flows (including precipitation, sewage, and stormwater); biodiversity (species diversity by taxa; habitat area), population (number of residents, employees, and visitors), economic costs of construction and demolition	this chapter
sECOM	New York/New Jersey Harbor (NYHOPS)	3-D x time	200–1,000 m	circulation, tides, storm surge, flooding	Georgas and Blumberg (2010)

* There may be slight differences in the definition of the spatial extent of domains even if they have the same focus (i.e., not all "Jamaica Bay" domains are identical.)

Table 8-1. Models are listed by domain (i.e., the area covered), dimensionality, resolution (i.e., smallest area for which data are calculated), outputs, and reference. Models are best developed for physical processes in the Jamaica Bay watershed, less so for ecological and social aspects of the system.

of the integrated modeling processes, then discuss Jamaica Bay–specific models within the physical, ecological, and socioeconomic domains. For each computational model, we summarize what it does, who uses it and why, and summarize in brief how it works. We close with some general observations for future directions that may be applicable in other situations.

Integrated Modeling Approaches to Resilience Planning

After the devastating 2005 hurricane season, which included two major hurricanes making landfall in the state of Louisiana (Katrina and Rita), a new Coastal Protection and Restoration Authority developed an integrated, participation-based decision process to develop a comprehensive master plan (Groves et al., 2014). This effort was designed to ensure that future coastal investment decisions actually address the scale of Louisiana's long-term coastal resilience challenges, as well as being well-coordinated and sustainable. Local communities and stakeholders played a significant role in the master planning process. This process resulted in a fifty-year, $50-billion integrated master plan that was passed unanimously by the state legislature in April 2012 and now guides all state-level coastal restoration and risk reduction investments. The disaster with the Deepwater Horizon oil platform—a disturbance of massive proportions—created funding streams that enabled Louisiana to take on resilience planning at a scale never before attempted in the United States.

A key feature of Louisiana's 2012 Coastal Master Plan analysis was a set of seven integrated computational models applied to consider future changes to the coastal system (Peyronnin et al., 2013). Key processes represented included ecohydrology, wetland morphology, barrier islands, vegetation, ecosystem services, storm surge and waves, and flood risk and damage. Each project and coastwide alternative proposed for the master plan was evaluated using this same suite of models, and was evaluated against the same set of future scenarios to estimate project benefits and rank them with respect to goals. The integrated models provided a novel science-based platform upon which investments in storm and flood risk reduction, land building, and ecosystem restoration could be compared objectively, highlighting high-performing projects, as well as key trade-offs. Using a separate planning tool (Groves and Sharon, 2013), computational model outputs were also used to support active deliberations, providing real-time analysis to resolve trade-offs and reach consensus.

Integrated modeling frameworks are becoming the norm for large, important estuarine systems. Knowles and Lucas (2015) developed the second generation of a large integrated modeling suite for the Sacramento–San Joaquin watershed, including climate modeling, hydrological linkages, hydrodynamics, phytoplankton dynamics, geomorphology, sediment supply, contaminants, and food web effects. Cloern et al. (2011) used an integrated framework to describe how the combined San Francisco Bay-Delta-River system might evolve with climate change over the next one hundred years. A similarly large-scale cooperative integrated modeling system has been developed for the Chesapeake Bay watershed (Christensen et al., 2009.)

Currently, a cooperative planning effort is under way in Jamaica Bay, supported by the Rockefeller Foundation and led by the RAND Corporation in collaboration with public agencies and researchers from the Science and Resilience Institute of Jamaica Bay (Fischbach and Knopman, pers. comm.) The expected outcomes of the study are to (1) identify key planning goals in Jamaica Bay from a wide variety of constituencies, including public agencies, local residents, and other stakeholders; (2) develop and test strategies to mitigate the most important future threats to these goals; (3) develop an understanding of key trade-offs among feasible investment paths across a range of scenarios; and (4) point the way toward a prioritized set of investment options. Future investment options could consist of different combinations of "green" and "gray" coastal infrastructure projects, such as wetlands restoration, barrier reefs, augmented beaches and dunes, revetments, flood walls, or surge barriers.

Integral to the planning process is the development of an integrated suite of simulation models for the Jamaica Bay watershed. These models will be applied to estimate future flood risk, water quality, and ecosystem outcomes in different scenarios, taking into account how the bay might change in a future without action, given sea level rise and other climate drivers. The integrated model will also be applied to estimate the benefits and costs of different proposed investments for the bay to highlight high-performing approaches in the face of future uncertainty and to identify key trade-offs to be resolved.

Visionmaker.nyc

Another approach to integrated modeling is the Visionmaker.nyc project (Sanderson, 2014; figures 8-1 through 8-4). Visionmaker is a free, Internet-based forum designed to enable anyone (land managers, politicians, neighborhood residents, schoolchildren, etc.) to develop and share climate-resilient designs for their own neighborhood. Visionmaker combines social media tools, environmental modeling, scenario analysis, and geographically explicit data to support participatory processes, acknowledge different stakeholders, and accrue community benefits from a transparent planning (Adger, 2003; Reed, 2008).

In Visionmaker, users create "visions" for neighborhoods of their own choosing. Visions are constructed of ecosystem/land use configurations, lifestyle choices, and climate scenarios. Note that ecosystems include "built" ecosystems, such as buildings, streets, and sidewalks, as well as forests, wetlands, and parklands. Ecosystems are of two types: base ecosystems, of which there can be only one type per 33-square-foot (10 m²) cell, and modifiers, of which there can be multiple types per cell. Base ecosystems refer

Figure 8-1. Splash screen for Visionmaker.nyc, a free, online, ecological democracy tool that incorporates models of environmental performance. Courtesy of Mario Giampieri of the Wildlife Conservation Society.

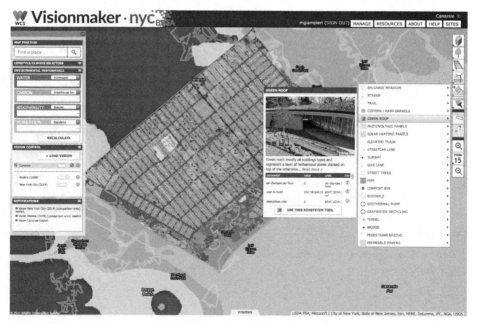

Figure 8-2. Visionmaker provides tools for users to "paint" ecosystems over existing neighborhoods and examine the changes in the environment. Ecosystems include built ones, such as buildings and roads, as well as natural ecosystems, such as salt marshes and beaches. Courtesy of Mario Giampieri of the Wildlife Conservation Society.

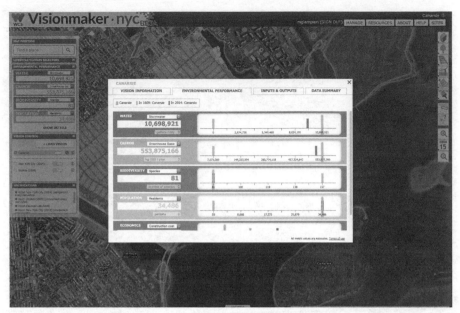

Figure 8-3. Visionmaker reports metrics of the water and carbon cycles, biodiversity, and population. Calculations are made for the user's vision in contrast to the same part of the city today and as it existed in the predevelopment state (based on data from the Welikia Project.) Courtesy of Mario Giampieri of the Wildlife Conservation Society.

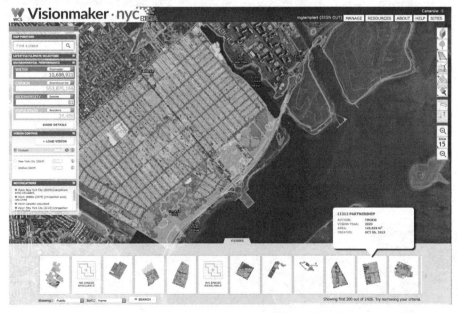

Figure 8-4. Visionmaker enables users to share their visions with others through the interface, enabling asynchronous collaboration. Additional functionality allows users to form groups to work on shared visions and challenges to designate areas of design and desired metrics of resilience. Courtesy of Mario Giampieri of the Wildlife Conservation Society.

to the main use type of the cell, including buildings, transportation types, or natural ecosystems. Modifiers change the ecosystem parameters of the cell, but do not replace the base ecosystem. For example, a green roof changes the biomass and water-holding capacity of an apartment building, but does not change the amount of floor area available for residential use. Similarly, street trees, bike lanes, trails in a forest, and piers in an estuary all model modifications, but not wholescale changes, to ecosystems.

Lifestyle choice options include parameterization for average American, average New Yorker, average earthling, average Lenape person—a Native American, representing a semihorticultural lifestyle as existed in the predevelopment New York City landscape—and an environmentally conscious, "eco-hipster" lifestyle. Climate scenarios draw on work from the New York City Panel on Climate Change (Horton et al., 2015). Visionmaker calculates metrics of environmental performance to compare the user's vision with the current condition based on modern geographic data describing the city today (color plate VI) and historical condition based on the Welikia Project (color plate I; Sanderson, 2009). Current models include a storm-event precipitation and runoff model (the "water" model); an energy, transportation, and ecosystem carbon cycling model (the "carbon" model); a species habitat and diversity model (the "biodiversity" model); a residential and worker model (the "population" model); and costs of demolition and construction (the "economics" model), all of which are explained briefly below. After a period of editing and revision, users may then choose to share their visions with selected others or with the public at large. Users may also issue "challenges" in terms of vision metrics.

Use of Computational Models in Jamaica Bay
In the next few sections we review individual models of the physical, ecological, and social systems specific to Jamaica Bay.

Models of Physical Systems
Computational models of the physical systems highlight the climatology, hydrology, oceanography, and sedimentology of Jamaica Bay.

Climate-Ocean Forecasts: Sea Level Rise and Coastal Storms
Global climate change will have local effects on Jamaica Bay. The New York City Panel on Climate Change 2 (NPCC2) made predictions for the future based on analysis of data from Coupled Model Intercomparison Project Phase 5 (CMIP5) global climate models (Taylor et al., 2011; Horton et al., 2015a, b). They estimated sea level rise to include the global effects of thermal expansion; loss of ice from Greenland and Antarctic ice sheets, glaciers, and ice caps; and changes in water storage on land. They also included local effects from

glacio-isostatic adjustments; gravitational, rotational, and so-called "elastic fingerprints" of ice loss on ocean levels in the New York City region, and included changes in ocean height associated with changes in local ocean density and circulation patterns. Scenarios of sea level rise, with measures of uncertainty, were produced for the 2020s, 2050s, 2080s, and 2100 relative to the base period of 2000–2004 (Horton et al., 2015b).

Future temperature and precipitation was modeled from the CMIP5 projections, accounting for interactions and feedbacks among chemistry, aerosols, vegetation, ice sheets, and biogeochemical cycles (Taylor et al., 2011). Some models include better treatments of rainfall and cloud formation that can occur at small "subgrid" spatial scales, and other improvements have led to better simulation of many climate dynamics. Local New York City results are derived from the "grid box" that covers the city and nearby Jamaica Bay. These spatial resolutions range from as fine as 50 miles by 40 miles (80 × 65 km) to as coarse as 195 square miles (315 km²), with an average resolution of approximately 125 miles by 115 miles (200 × 185 km). The combination of thirty-five global climate models and two representative concentration pathways produces a 70-member (35 × 2) matrix of outputs for temperature and precipitation (Horton et al., 2015a).

The NPCC2 computed results for future time periods that were compared to the climate model results for the baseline period (1971–2000). Mean temperature change projections are calculated via the delta method, a bias correction using the difference between each model's future and baseline simulation, rather than direct model outputs. The delta method was previously used for local climate change projections in New York (Horton et al., 2011). Mean precipitation change is also estimated by taking the ratio of a model's future precipitation projections compared to baseline precipitation (expressed as a percentage change). As with sea level rise, the NPCC2 developed scenarios of temperature and precipitation in the future: 2020, 2050, 2080, and 2100 (Horton et al., 2015a).

Precipitation

Precipitation falls on the Jamaica Bay watershed in the form of rain and snow. New York City has a combined sewer system that directs stormwater drainage into the same pipes as the sanitary sewage flows from toilets, showers, etc. The New York City Department of Environmental Protection (NYCDEP) uses an integrated water modeling system called InfoWorks to manage both of these aspects of the combined sewer flows (NYCDEP, 2012). The InfoWorks models calculate runoff using methods originally contained in the U.S. Environmental Protection Agency's Storm Water Management Model (EPA SWMM; Rossman, 2015). In these models, when rain falls on pervious surfaces, a fraction infiltrates into the soil; the remainder of the water plus runoff from impervious surfaces becomes overland surface flow (i.e., runoff) that is then routed to the entry point to a storm or combined sewer. InfoWorks uses a Horton infiltration equation based on the hydrological characteristics of the soil to calculate the amount of rainfall that infiltrates into the ground from pervious surfaces. Rain falling on an impervious surface is subject to a small loss through ponding on the surface, with the remainder of the flow becoming overland flow and runoff into the sewer system. Water may also be transported to the atmosphere via evapotranspiration that is estimated on a monthly basis from observational data. The InfoWorks hydrological model has been calibrated by adjusting estimates of impervious area on a site-specific basis such that model estimates match observed flows into each of the city's wastewater treatment plants.

A simpler, less highly tuned model of hydrological flows was developed as part of the Visionmaker.nyc platform. The Visionmaker water model is based on a simple, flow-through storm-event model, adapted from Vörösmarty et al. (1996) and Mitchell et al. (2001). The model estimates piped water demand based on an estimated population within the vision extent and the lifestyle of the people living there. Piped water passes through to become the sanitary sewage flow. Exterior water flows come from precipitation, which is a function of the user's selected storm-event (assumed to occur in June) and climate scenario. Evapotranspiration is estimated using the Hamon method (Hamon, 1961). Precipitation is distributed proportionally by area to impervious and pervious surfaces and water that exceeds water storage parameters flows into the storm drain system. Stormwater drainage capacity assumes that built ecosystems are constructed to New York City design standards. Flows beyond the capacity of the storm drain system flow to streams (if present) or are tabulated as flooding within the vision extent.

Groundwater

Misut and Voss (2004; 2007) parameterized a three-dimensional version of the U.S. Geological Survey's (USGS) Saturated-Unsaturated Transport (Sutra) model to estimate groundwater flow and the movement of the freshwater/saltwater interface on western Long Island and into Jamaica Bay. Sutra models fluid-density-dependent groundwater flows. Model bottom (i.e., bedrock) was simulated as impermeable, and a detailed representation of the upper glacial, Jameco, Magothy, and Lloyd aquifers (see chapter 4) is presented in terms of the depths of each layer and the vertical and horizontal hydraulic conductivities. Saltwater hydrostatic pressure boundaries were applied in offshore zones, and freshwater hydrostatic pressure boundaries were applied at streams, ponds, and along the eastern boundary of the model, which represents the continuation of aquifers eastward under Long Island. Misut and Voss (2007) used this model system to test the regional effects of using the aquifer systems for emergency water storage and recovery. Previously, a four-layer regional groundwater flow model (MODFLOW; McDonald and Harbaugh, 1988) was developed for Kings and Queens Counties, New York, as described by Buxton and Shernoff (1995), Kontis (1999), and Misut and Monti (1999).

Circulation and Storm Surge

Several research groups have created hydrodynamic models of the bay, ranging from two-dimensional (2-D) flood modeling to three-dimensional (3-D) circulation modeling, and from relatively coarse scales of hundreds of meters to fine scales below 33 feet (10 m) (e.g., table 8-1) to study patterns of circulation and storm surge flooding.

One model is the Stevens Institute Estuarine and Coastal Ocean hydrodynamic Model (sECOM) (e.g., Georgas and Blumberg, 2010; Blumberg et al., 1999). Two-dimensional flood modeling was performed for flood adaptation studies on a 98-foot (30 m) model grid (figure 8-5) that extends upland to 20 feet (6 m) land elevation (Orton et al., 2015a), nested inside the regional grid of the New York Harbor Observing and Prediction System (NYHOPS; e.g., Georgas et al., 2014). More advanced 3-D modeling, wave modeling, and vegetation-flow interactions (Marsooli et al., 2016a, b) are also being developed. Operational forecasts of Jamaica Bay circulation and waves, and ensemble (probabilistic) forecasts of water levels are also available from the NYHOPS system (http://stevens.edu/maritimeforecast), though resolution is low in Jamaica Bay (656–3,280 feet [200–1,000 m]). Resolution elsewhere in New York Harbor is typically about 328 feet (100 m), and the NYHOPS forecast system has been demonstrated to have high accuracy across the region (Georgas and Blumberg, 2010).

Flood, wave, and flood adaptation simulations were recently conducted using the ADCIRC/SWAN coupled model (Dietrich et al., 2011) on grids of up to 131-foot (40 m) resolution for studies of flood zones with added sea level rise (Orton et al., 2015b), as well as coastal flood adaptation, including Jamaica Bay. ADCIRC is an advanced ocean circulation model (Westerink et al., 1993), typically run in 2-D, but also with 3-D capabilities; SWAN is a nearshore wave model (Booij et al., 1999). The flood adaptation work with both ADCIRC/SWAN and sECOM on the 98-foot (30 m) grid showed that wetlands in the center of the bay have little effect on flood elevations, though they may help reduce wave heights and erosion during storms (Orton et al., 2015a).

Sediments

Wilson and Flagg (in prep.; also see Wilson, 2008) described detailed 3-D hydrodynamic and sediment transport modeling that was performed with the FVCOM model (Finite-Volume Coastal Ocean Model; Chen et al., 2007). FVCOM describes basic tidal and estuarine hydrodynamics to simulate movement of fine-grain and coarse sediments (figure 8-6), including tidal characteristics and asymmetries in response to the bathymetry and estuarine circulation and stratification, and wetting and drying of marshes. FVCOM has the flexibility to increase resolution in particular areas of interest in Jamaica Bay, where bathymetric gradients may be steepest. The model incorporates recent multibeam bathymetry collected by Flood (2011). Simulations were validated

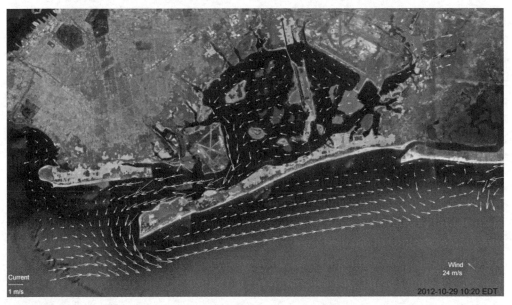

Figure 8-5. Stevens ECOM (sECOM) simulation of Hurricane Sandy floodwater depths (shading) over ground, along with vectors for water velocity (maximum shown, about 1 m/s). This snapshot of the model simulation shows the time of peak flood depths in neighborhoods surrounding the bay, as the flow was beginning to recede out of the inlet. Courtesy of Philip Orton of the Stevens Institute of Technology.

against available information on tide range and against available information on mean salinity (Wilson, 2008).

Integrated Physical Model

A team with members at the U.S. Geological Survey and Louisiana State University (Hongqing Wang and others), working with Philip Orton at the Stevens Institute of Technology and collaborators at the City of New York's Department of Parks and Recreation and the National Park Service, is currently developing an integrated numerical modeling system that couples wind, waves, storm surge, sediment transport, hydrodynamics, and wetland morphologic dynamics on the Delft3D platform (Orton, pers. comm.) The researchers are currently developing the new integrated model, and are applying this model to hindcast Hurricane Sandy and investigate the poorly understood long-term effects of major storm events on sediment movement and dynamics in the bay.

Ecological Models

Computational models of ecological interactions simulate living systems in combination with physical aspects of Jamaica Bay. Existing approaches include models of

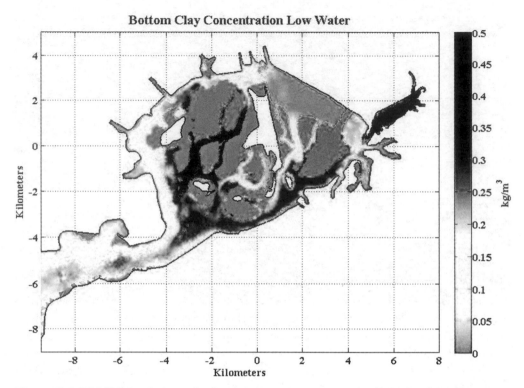

Figure 8-6. FVCOM simulations of sediment movement associated with both tidal and estuarine circulation. The distribution of near-bottom clay concentration at low water shows evidence for resuspension and transport in Pumpkin Patch, Big Fishkill, and Beach Channels (from Wilson and Flagg, in prep). Courtesy of Bob Wilson of Stony Brook University.

water quality, carbon cycling, ecosystem distributions, and species and species habitat distributions.

Water Quality

The NYCDEP uses a set of calibrated mathematical models to describe relationships between combined sewer and stormwater discharges to Jamaica Bay and the water quality in the water body. InfoWorks (described above) was used to calculate the flows and loadings of pollutants that are fed to the receiving water models. The Jamaica Bay Eutrophication Model (JEM), a 3-D, time-variable hydrodynamic and water quality model containing a twenty-eight-state variable eutrophication model for computing nutrient forms and chlorophyll a (algae) concentrations provided the basis of the water quality modeling analysis (HydroQual, 2002).

The hydrodynamic component of the JEM model uses the volume and velocity information from effluent lines, along with additional inputs and kinetic equations, to calculate receiving water concentrations for different types of pollutants. The model includes a sediment nutrient flux submodel that calculates organic decomposition within the sediment and the flux of inorganic materials between the sediment and overlying water column. JEM was calibrated during the Jamaica Bay Eutrophication Study (HydroQual, 2002) and was peer reviewed (HydroQual, 2003). The original JEM grid is relatively coarse by today's standards, but was considered to be a moderate to high-resolution grid in the mid-1990s. The 1995/1996 field program included routine water chemistry and dissolved oxygen measurements as well as sediment oxygen demand, nutrient fluxes, measurements of macroalgal biomass and nutrient content, wetland biomass and nutrient content and biomass estimates of the hard-shell clam, *Mercenaria mercenaria*. The JEM water quality model, which included a sediment diagenesis/nutrient flux submodel (DiToro and Fitzpatrick, 1993; DiToro, 2001) and a suspension feeder submodel (Cerco and Meyers, 2000; HydroQual, 2000), used data from these studies to calibrate and validate the hydrodynamic and water quality models. The model was subsequently used by NYCDEP to help develop a nitrogen control plan for the bay to reduce eutrophication within the bay (Isleib and Fitzpatrick, 2007; NYCDEP, 2015).

Recently, HDR/HydroQual (2012) developed a revised JEM grid with a higher degree of spatial resolution, so as to be used in a study of oyster larvae transport and survival in Jamaica Bay. A particle tracking model, with biological behavior (based on a framework developed by North et al. [2008]) was used to explore optimal locations for placing oyster reefs. Under National Park Service funding (2014–2016), additional improvements are being tested, including an ulva biomass submodel, vegetation drag on water flow (e.g., Marsooli et al., 2016a, b), recent bathymetric data from USGS and Stony Brook, and updated model parameters based on recent science and field observations. These possible improvements raise the possibility of finalizing an improved version of JEM by the end of 2016 (e.g., JEM2016).

Carbon Cycling

Visionmaker includes a carbon model that includes two distinct aspects of the carbon cycle; the movement of carbon from the atmosphere via net primary productivity, and carbon emissions from consumption of fossil fuels. The ecosystem carbon cycling includes estimates of average net primary productivity for each ecosystem in the user's vision on annual basis. Carbon is followed through litter fall and decomposition into soil organic matter. Respiration is estimated for animals (people, pets, and wildlife) and the soil. Food and food waste are also estimated based on the population and lifestyle of the user's vision.

Submodels also simulate energy consumption by buildings and transportation. The energy model is based in part on review standards laid out in the New York City Environmental Quality Review. It includes a climate-dependent estimate of energy used for heating and cooling based on a degree-days approach and the area of exterior roof and wall. Buildings are described in terms of human uses, which in turn result in energy consumption for residential and business use cases. Electricity can be supplied from the grid or other distributed energy generation methods (such as solar panels.) Transportation energy use is based on a trip generation calculator. Trips are broken into distances traveled by mode. All of the energy consumption estimates are eventually translated into fuel consumption amounts, which are then used to estimate carbon dioxide and methane emissions.

Ecosystems

Sanderson (2009) derived a set of conditional relationships representing the distributional requirements of New York City relevant ecological communities described in Edinger et al. (2014). Called "Muir webs" (after the naturalist John Muir), these network-based relationships express qualitatively the environmental conditions where ecosystems are mostly likely to occur in the New York City region, including factors such as topography, soil type, frequency and depth of flooding, wave action, and fire. By combining maps of these factors in a geographic information system (GIS) framework, the spatial distribution of the ecological communities can be estimated. These methods are currently being used in the Welikia Project (welikia.org) to estimate the predevelopment distribution of species in the Jamaica Bay watershed based on sea level, topography, bathymetry, soil type, and other factors.

Two other models focus on the evolution of salt marsh habitat that are applicable to Jamaica Bay and have been proposed for application in the RAND integrated modeling plan detailed above. The Sea Level Affecting Marshes Model (SLAMM) simulates the "dominant processes in wetlands conversions and shoreline modifications during long-term sea level rise" (USFWS, 2011). The five processes modeled include inundation, erosion, overwash, saturation, and accretion. SLAMM makes use of existing public data sets such as the U.S. Fish and Wildlife Service's National Wetlands Inventory, U.S. Geological Survey Digital Elevation Models, and others. It can be run in either 2-D or 3-D configurations, each using equal cell sizes that are assigned to a specific coastal land class. The New York City Department of Parks and Recreation and The Nature Conservancy are currently working on an application of SLAMM to New York City marshes (Marit Larsen, pers. comm.)

A second salt marsh change model, the Marsh Equilibrium Model, forecasts changes in marsh elevation and marsh productivity based on inputs related to marsh biomass, tidal changes, rate of sea level rise, sediment budget, and other factors (Morris, 2015).

Tidal changes can be modeled in hydrodynamic models such as the ones discussed above (e.g., ADCIRC, sECOM, or Delft3D). Marsh equilibrium outputs are calculated at each cell, as a function of the hydrodynamics; and over time, changes in marsh heights and roughness can be fed back to the hydrodynamic model, creating a two-way coupling to simulate long-term change.

Species

Although there have been surveys of species in Jamaica Bay (summarized in chapter 5), species distributional modeling for the bay is in its infancy. The same Muir web approach described by Sanderson (2009) is being extended to Jamaica Bay species as part of the Welikia Project. Muir webs document species requirements by working recursively from textual descriptions of species requirements as shown in a field guide or autecological study (Sanderson, 2009). A Muir web element may be a species (e.g., *Spartina alterniflora*), an abiotic factor (elevation), or process (tidal flooding). The Muir web map of a species distribution represents the areas where all the requirements for the species are met.

An alternative method is being developed for the Visionmaker project where the "area of ecosystems" is taken as a given for the vision. Meixler et al. (2015) developed a matrix of species-habitat relationships for Jamaica Bay. The area of a species's habitat in a vision is based on the sum of the areas of the relevant ecosystem types. The area of habitat is then compared with a species-specific minimum habitat area size. For plants, this minimum is based on a circle with radius equal to the average seed distribution distance. For animals, it is based on a home range size estimate. This rapidly computable method will enable us to estimate a potential species list for a given vision.

Biodiversity

Visionmaker also includes a simple biodiversity model that predicts the number of species by taxa (i.e., plants, birds, mammals, amphibians, reptiles, fish) based on species-area relationships. Habitat areas are estimated based on ecosystem distributions; for the user's vision, Visionmaker estimates the species diversity of the scenario in contrast to that same part of the city today and to the predevelopment state. Ongoing work is designed to add species-specific models for species of concern in Jamaica Bay. These simple models will document which ecosystems certain Jamaica Bay species use as habitat, and simple minimum size thresholds for the amount of area required for a population.

Socioeconomic Models

Computational socioeconomic models use equations to simulate the dynamics of what people do. Like the models above, they can be connected into larger sets of integrated models that allow one to examine how human decision making affects the physical or ecological systems, or vice versa. In physical and ecological models, human decision making is often treated as endogenous inputs; for example, the construction of a seawall affects the tides or land fill influences species distributions. But socioeconomic models could also suggest how we make decisions about resilience.

Overall, this area of research is less developed for Jamaica Bay than the physical and ecological models reviewed above. What work exists is mainly private, conducted by companies who have a proprietary interest in model methods and results. Although the city, state, and federal agencies (such as the New York City Economic Development Corporation and Department of City Planning, the U.S. Department of Housing and Urban Development) actively conduct a variety of projects, they rarely have specific models designed for Jamaica Bay communities.

Damage to Structures from Flood

One of the broadly shared values for Jamaica Bay is to reduce damage from future flooding events (as discussed in chapters 2 and 3). Flooding as a social problem requires integrating hydrodynamic models (as described above) with estimates of social vulnerability (e.g., Aerts et al., 2013). For example, the U.S. Army Corps of Engineers (USACE) Hydrological Engineering Center has developed the Flood Damage Reduction Analysis (HEC-FDA) software developed by the USACE Hydrologic Engineering Center (http://www.hec.usace .army.mil/software/hec-fda/). David Miller and Associates is applying this model in support of the USACE Jamaica Bay Reformulation Study (Knopman and Fischbach, pers. comm.). HEC-FDA computes expected annual damages from flooding and annual exceedance probabilities and conditional nonexceedance probabilities that are used in levee certification.

The Federal Emergency Management Agency has developed the HAZUS-MH Flood Model, a GIS-based flood hazard modeling tool (https://www.fema.gov/protecting-our -communities/hazus-mh-flood-model). Like HEC-FDA, HAZUS-MH enables one to study annualized losses and test scenarios of flood protection. It includes two modules: a flood hazard analysis and a flood loss estimate analysis. The hazard module calculates flood depth, elevation and flow velocity from data on the frequency, water discharge, and ground elevations. The flood loss translates the flood hazard information into estimates of physical damage and economic loss.

As part of the Louisiana integrated resilience planning effort described above, the RAND Corporation developed the Coastal Louisiana Risk Assessment (CLARA) damage module framework (Johnson et al., 2013). Similar to the HAZUS-MH and HEC-FDA, CLARA uses asset inventory data and flood depths to estimate direct damage from flooding. A similar framework could support damage assessment for individual tax parcels in Jamaica Bay (Fischbach and Knopman, pers. comm.)

Population Growth

Understanding population change is a critical factor in understanding resilience in Jamaica Bay or other urbanized estuaries, yet the actual methods currently used to estimate population relative to other inputs are quite simple. In New York City, as in other parts of the United States, the decennial census provides population enumerations at a high spatial resolution. The New York City Department of City Planning (Salvo et al., 2006) created a baseline demographic model by estimating and extrapolating series of fertility, mortality, and migration rates by age/sex, calibrating projections to census data, and then projecting forward in time assuming similar demographic rates in the future. They adjust for reported undercounts of population in the city. Currently estimates are made at the scale of city boroughs, though presumably the methods could be specified for Jamaica Bay neighborhoods.

Visionmaker has a simple space-based population model that estimates the area of residential floor space based on the distribution of ecosystems within a vision. A residential population density is estimated for each of the five lifestyles, representing the amount of residential floor space occupied by one human being on average. Vision population estimates are adjusted for the residential vacancy rates (which tend to be quite low in New York City.) Similar population estimates are made for the worker population based on the area of office, retail, and other employment-based floor areas within the vision, and for the visitor (i.e., tourist) population, based on the floor area of hotels. Population estimates are used to drive estimates of water and energy consumption and trip demand.

Transportation Demand

The New York City Metropolitan Transportation Council has developed a Best Practices Model (BPM) to estimate transportation demand and flows in the New York City metropolitan region, including around Jamaica Bay (http://www.nymtc.org/project/bpm/bpmindex.html). The BPM was constructed in the 1990s in response to requirements of the Intermodal Surface Transportation Efficiency Act and the Clean Air Act. The BPM models a thirty-one-county area as 3,586 discrete transportation analysis zones using microsimulation, where travel by individuals is estimated individually. The model consists of a "household, auto-ownership and journey-frequency" submodel that generates a number of trips, and then a "mode destination stop choice" submodel that assigns trips to transportation modes and distances, including intermediate destinations between home and work (or final destination). Household interviews conducted in 1996–1997 were used along with speed, traffic counts, and cordon counting data to estimate the parameters. The highway network is modeled as 52,794 links including all minor arterial and larger roadways, and a transit route system that consists of commuter rail, bus, subways, ferries, and other modes of shared transportation.

Visionmaker has a much simpler transportation model. Within a vision, trip demand is a function of the floor area for different use types (i.e., residences, office, manufacturing, public assembly) and lifestyle, with separate estimates for working and nonworking days. Lifestyle also determines the proportion of trips by mode and distance, where distance is represented by a small set of categories (e.g., 0–0.25 mile, 0.25–1 mile [0–0.4 km, 0.4–1.6 km], etc.). The number of trips and distance for each mode is then used to estimate the amount of fuel consumed for each mode, and the fuel types, which then feeds into the carbon model. Freight trips are estimated using a freight trip generation rate density that varies by use type. Lifestyle determines the proportion of freight trips by model. An average freight shipment distance by lifestyle and mode leads to an estimate of the distance traveled, and therefore the fuels consumed, to deliver freight into and out of the vision.

Economic Costs of Construction and Demolition

Visionmaker has a simple construction and demolition cost metric associated with changes in the ecosystem types within visions. When base ecosystems are changed, Visionmaker makes an estimate of the cost of the change in terms of 2014 dollars. Previous ecosystems are assumed to be demolished, and the new ecosystems constructed, where "construction" can also mean "restoration" in terms of natural ecosystems. Costs of demolition and construction are parameterized on a per area basis from a review of the literature (e.g., RSmeans).

Gaps in Knowledge and Recommendations for Future

To guide future efforts, we have a small list of recommendations. These recommendations could apply to other urban estuaries as well as Jamaica Bay.

Advance Model Integration

Ensuring a system's resilience, as was stressed in chapter 2, requires an integrated approach that connects the physical, ecological, and social systems of site. As described in this chapter, many models are currently applied to evaluate different portions of the coastal system, but better integrated systems are needed to allow these model processes to feed one another and for proposed interventions to be tested in a common set of simulation models. The work in Louisiana after Hurricanes Rita and Katrina, as well as work in other estuaries, provides a template for how integration can occur.

Invest in Better Models of Ecological and Socioeconomic Systems

More work needs to be emphasized in creating models that treat and integrate ecological and social aspects of the system, including population change, land use change, and resulting consequences for how the bay is used by different groups of people, building off of studies such as those described in chapter 3. We need to add economics into our understanding of processes supporting resilience, because most decisions about development are based in economic decision making, and connect economic decision making to ecological and physical outcomes. We need to do a better job collecting the data to validate models of ecological and social behaviors (see reviews in chapters 5 and 6; also Campbell et al., 2014), and use that data to drive model creation and use. As we have seen, most models today respond to regulatory demands or the shock that follows damaging events (e.g., Hurricane Sandy), rather than letting investigations into the values for, and drivers of, resilience drive the pace and funding for modeling efforts.

The Visionmaker project, while itself in its infancy, presents the possibility of using the Internet to give a much wider variety of people an opportunity to use computational models to enhance their own understanding and visions for resilience in Jamaica Bay, increasing participation in our collective efforts to build resilience, integrating across physical, ecological, and social subsystems.

Create Open-Source Models

Computational models require collaboration across disciplines to be of greatest value for resilience planning and testing. Those collaborations are facilitated by open-source approaches, where parameters, model relationships, and even computer code are available for review and use by other modeling teams (Pearce, 2012). In other domains, great advances have been made through free and open-source software (Wheeler, 2015; Woelfle et al., 2011), and the Internet provides opportunities for model integration and intercommunication. For Jamaica Bay, and other coastal areas being studied for resilience capacity, we need not only models that are published and in the public domain, but mechanisms to generate scenarios that address resilience from various perspectives. Resilience is ultimately a matter of choosing what aspects of the system society values the most (see chapters 2, 3), so the models that we build and choose must be responsive to aspects of the system we hope to make resilient.

Acknowledgments: The authors acknowledge the contributions of James Fitzpatrick, Scott Hagen, and James Morris to an earlier draft of this chapter.

References

Adger, W.N. 2003. Social capital, collective action, and adaptation to climate change. *Economic Geography* 79: 387–404. doi:10.1111/j.1944-8287.2003.tb00220.x.

Aerts, J.C., Lin, N., Botzen, W., Emanuel, K., and de Moel, H. 2013. Low-Probability flood risk modeling for New York City. *Risk Analysis* 33: 772–788. doi:10.1111/risa.12008.

Alcamo, J. 2008. *Environmental Futures: The Practice of Environmental Scenario Analysis.* Amsterdam: Elsevier.

Blumberg, A.F., Khan, L.A., and St. John, J.P. 1999. Three-dimensional hydrodynamic model of New York Harbor region. *Journal of Hydrological Engineering* 125(8): 799–816.

Booij, N., Ris, R.C., and Holthuijsen, L.H. 1999. A third-generation wave model for coastal regions, Part I, model description and validation. *Journal of Geophysical Research* 104: 7649–7666.

Buxton, H.T., and Shernoff, P.K. 1995. Ground-water resources of Kings and Queens Counties, Long Island, New York (Open-File Report 92-76). U.S. Geological Survey, Albany, NY.

Campbell, L., Svendsen, E., Falxa-Raymond, N., and Baine, G. 2014. Reading the landscape, a reflection on method. *PLOT* 3: 90–95.

Cerco, C.F., and Meyers, M. 2000. Tributary refinements to Chesapeake Bay model. *Journal of Environmental Engineering* 126(2): 164–174.

Chen, C., Huang, H., Beardsley, R.C., Liu, H., Xu, Q., and Cowles, G. 2007. A finite volume numerical approach for coastal ocean circulation studies: Comparisons with finite difference models. *Journal of Geophysical Research* 112, C03018. doi:10.1029/2006JC003485.

Christensen, V., Beattie, A., Buchanan, C., Ma, H., Martell, S.J., et al. 2009. Fisheries Ecosystem Model of the Chesapeake Bay: Methodology, Parameterization, And Model Exploration. US Department of Commerce, National Oceanic and Atmospheric Administration, National Marine Fisheries Service.

City of New York. 2013. A Stronger, More Resilient New York. Mayor's Office of Long-Term Planning and Sustainability, New York.

Cloern, J.E., Knowles, N., Brown, L.R., Cayan, D., Dettinger, M.D., et al. 2011. Projected evolution of California's San Francisco Bay-Delta-River system in a century of climate change. *PLoS ONE* 6, e24465. doi:10.1371/journal.pone.0024465.

Dietrich, J.C., Zijlema, M., Westerink, J.J., Holthuijsen, L.H., Dawson, C., et al. 2011. Modeling hurricane waves and storm surge using integrally coupled, scalable computations. *Coastal Engineering* 58: 45–65. doi:10.1016/j.coastaleng.2010.08.001.

DiToro, D.M. 2001. *Sediment Flux Modeling*. New York, NY: John Wiley & Sons.

DiToro, D.M., and Fitzpatrick, J.J. 1993. Chesapeake Bay Sediment Flux Model. Tech. Report EL-93, US Army Corps of Engineers Waterways Experiment Station.

Edinger, G.J., Evans, D.J., Gebauer, S., Howard, T.G., Hunt, D.M., and Olivero, A.M. (eds). 2014. *Ecological Communities of New York State*. Second Edition. A revised and expanded edition of Carol Reschke's *Ecological Communities of New York State*. New York State Department of Environmental Conservation, New York.

Federal Emergency Management Agency, 2014. Region II Coastal Storm Surge Study Overview. Federal Emergency Management Agency, Washington, D.C.

Fischbach, J.R., Lempert, R.J., Molina-Perez, E., Tariq, A.A., Finucane, M.L., Hoss, F. 2015. *Managing Water Quality in the Face of Uncertainty: A Robust Decision Making Demonstration for EPA's National Water Program*. Santa Monica, CA: RAND Corporation. http://www.rand.org/pubs/research_reports/RR720.html.

Flood, R.D. 2011. High-Resolution Bathymetric and Backscatter Mapping in Jamaica Bay: Final Report to the National Park Service. Stony Brook University, Stony Brook, NY.

Gallopin, G. 2002. Planning for resilience: Scenarios, surprises, and branch points. In: *Panarchy: Understanding Transformations in Human and Natural Systems*. Gunderson, L.H., and Holling, C.S. (eds). Washington, D.C.: Island Press.

Georgas, N., and Blumberg, A.F. 2010. Establishing confidence in marine forecast systems:

The design and skill assessment of the New York Harbor Observation and Prediction System, Version 3 (NYHOPS v3), paper presented at Eleventh International Conference in Estuarine and Coastal Modeling (ECM11), ASCE, Seattle, Washington.

Georgas, N., Orton, P., Blumberg, A., Cohen, L., Zarrilli, D., and Yin, L. 2014. The impact of tidal phase on Hurricane Sandy's flooding around New York City and Long Island Sound. *Journal of Extreme Events*. doi:10.1142/S2345737614500067.

Groves, D.G., and Lempert, R.J. 2007. A new analytic method for finding policy-relevant scenarios. *Global Environmental Change Part A: Human & Policy Dimensions*, Vol. 17(1): 73–85.

Groves, D.G., and Sharon, C. 2013. Planning tool to support planning the future of coastal Louisiana. In: Louisiana's 2012 Coastal Master Plan Technical Analysis, *Journal of Coastal Research*, Special Issue No. 67, 1–15. Peyronnin, N., and Reed, D. (eds). Coconut Creek (Florida).

Groves, D.G., Fischbach, J.R., Knopman, D., Johnson, D.R., and Giglio, K. 2014. *Strengthening Coastal Planning: How Coastal Regions Could Benefit from Louisiana's Planning and Analysis Framework*. Santa Monica, CA: RAND Corporation. http://www .rand.org/pubs/research_reports/RR437.html.

Hamon, W.R. 1961. Estimating potential evapotranspiration. *Journal of the Hydraulics Division, Proceedings of the American Society of Civil Engineers* 87: 107–120.

Hawes, C., and Reed, C. 2006. Theoretical steps towards modelling resilience in complex systems. In: *Computational Science and Its Applications—ICCSA 2006*. Gavrilova, M., Gervasi, O., Kumar, V., Tan, C.J.K., Taniar, D., et al. (eds). 644–653. Berlin: Springer.

Horton, R.M., Gornitz, V., Bader, D.A., et al. 2011. Climate hazard assessment for stakeholder adaptation planning in New York City. *Journal of Applied Meteorology and Climatology* 50: 2247–2266.

Horton, R., Bader, D., Kushnir, Y., Little, C., Blake, R., and Rosenzweig, C. 2015a. New York City Panel on Climate Change 2015 Report, Chapter 1: Climate Observations and Projections. *Annals of New York Academy of Sciences* 1336: 18–35. doi:10.1111/nyas.12586.

Horton, R., Little, C., Gornitz, V., Bader, D., and Oppenheimer, M. 2015b. New York City Panel on Climate Change 2015 Report, Chapter 2: Sea Level Rise and Coastal Storms. *Annals of New York Academy of Sciences* 1336: 36–44. doi:10.1111/nyas.12593.

HydroQual. 2000. Development of a suspension feeding and deposit feeding benthos model for Chesapeake Bay. Prepared for the U.S. Army Engineer Waterways Experiment Station, Vicksburg, MS.

HydroQual. 2002. A Water Quality Model for Jamaica Bay: Calibration of the Jamaica Bay Eutropication Model (JEM), Hydroqual, Inc., Mahway, NJ.

HydroQual. 2003. Model Evaluation Group (MEG) Peer Review of the Jamaica Bay Eutrophication Model (JEM). Under subcontract to O'Brien and Gere Engineers, Inc. For the City of New York, Department of Environmental Protection. New York.

HydroQual. 2012. Defining Restoration Objectives and Design Criteria for Self-Sustaining Oyster Reefs in Jamaica Bay, HDR/Hydroqual, Mahway, NJ.

Isleib, R.R.P.E., and Fitzpatrick, J.J. 2007. The development of a nitrogen control plan for a highly urbanized tidal embayment. *Proceedings of the Water Environment Federation*, 296–320. doi:10.2175/193864707786619152.

Johnson, D.R., Fischbach, J.R., and Ortiz, D.S. 2013. Estimating surge-based flood risk with the Coastal Louisiana Risk Assessment Model. *Journal of Coastal Research* 2013/07/01, 109–126.

Kontis, A.L. 1999. Simulation of freshwater-saltwater interfaces in the Brooklyn-Queens aquifer system, Long Island, New York (USGS Numbered Series No. 98-4067), Water-Resources Investigations Report. U.S. Geological Survey, Reston, VA.

Knowles, N., and Lucas, L. 2015. CASCAaDE II Project Final Report: Computational Assessments of Scenarios for Change for the Delta Ecosystem. U.S. Geological Survey, Menlo Park, CA.

Marsooli, R., Orton, P.M., Georgas, N., and Blumberg, A.F. 2016a. Three-dimensional hydrodynamic modeling of storm tide mitigation by coastal wetlands. *Coastal Engineering* 111: 83–94.

Marsooli, R., Orton, P.M., Fitzpatrick, J., Georgas, N. and Blumberg, A.F. 2016b. Impact of salt marshes on residence time in Jamaica Bay, NY. Poster presented at the State of the Bay Symposium, June 16.

McDonald, M.G., and Harbaugh, A.W. 1988. *A Modular Three-Dimensional Finite-Difference Ground-Water Flow Model: U.S. Geological Survey Techniques of Water-Resources Investigations, book 6*, chap. A1.

Meixler, M.S., Sanderson, E.W., Fisher, K., Newton, E., and Sacatelli, R. 2015. Modeling biodiversity of the New York coastal urban fringe. Ecological Society of America. Baltimore, MD. August 14.

Misut, P.E., and Monti, J., Jr. 1999. Simulation of Ground-Water Flow and Pumpage in Kings and Queens Counties, Long Island, New York (Water-Resources Investigations Report 98-4071). U.S. Geological Survey, Coram, New York.

Misut, P.E., and Voss, C.I., 2004. Simulation of subsea discharge to Jamaica Bay in New York City with a three-dimensional, variable density, finite-element model. In: *International Conference on Finite Element Models, MODFLOW, and More: Solving Groundwater Problems.* Kovar, K., Hrkal, Z., and Bruthans, J. (eds). Czech Republic: Karlovy Vary.

Misut, P.E., and Voss, C.I., 2007. Freshwater–saltwater transition zone movement during aquifer storage and recovery cycles in Brooklyn and Queens, New York City, USA. *Journal of Hydrology* 337: 87–103. doi:10.1016/j.jhydrol.2007.01.035.

Mitchell, V., Mein, R., and McMahon, T., 2001. Modelling the urban water cycle. *Environmental Modelling and Software* 16: 615–629.

Morris, J. 2015. The Marsh Equilibrium Model MEM 3.4. http://129.252.139.114/model/marsh/mem2.asp.

New York City Department of Environmental Protection (NYCDEP). 2012. Infoworks: Citywide Recalibration Report. Updates to and Recalibration of October 2007 Landside Models. City of New York, New York.

NYCDEP. 2015. Nitrogen Control Program. http://www.nyc.gov/html/dep/html/harborwater/nitrogen.shtml.

New York Metropolitan Transportation Council, 2016. New York Best Practice Model (NYBPM). https://www.nymtc.org/Data-and-Modeling/New-York-Best-Practice-Model-NYBPM. Accessed 6/14/16.

North, E., Schlag, Z., Hood, R., Li, M., Zhong, L., et al. 2008. Vertical swimming behavior influences the dispersal of simulated oyster larvae in a coupled particle-tracking and hydrodynamic model of Chesapeake Bay. *Marine Ecology—Progress Series* 359: 99.

Orton, P.M., Talke, S.A., Jay, D.A., Yin, L., Blumberg, A.F., et al. 2015a. Channel shallowing as mitigation of coastal flooding. *Journal of Marine Science and Engineering* 3(3): 654–673. doi:10.3390/jmse3030654.

Orton, P., Vinogradov, S., Georgas, N., Blumberg, A., Lin, N., et al. 2015b. New York City

Panel on Climate Change 2015 Report, Chapter 4: Dynamic coastal flood modeling, *Annals of New York Academy of Sciences* 1336(1): 56–66.

Pearce, J.M. 2012. Open source research in sustainability. *Sustainability: The Journal of Record* 5: 238–243. doi:10.1089/SUS.2012.9944.

Peyronnin, N., Green, M., Richards, C.P., Owens, A., Reed, D., et al. 2013. Louisiana's 2012 Coastal Master Plan: Overview of a science-based and publicly informed decision-making process. In: *Louisiana's 2012 Coastal Master Plan Technical Analysis, Journal of Coastal Research,* Special Issue No. 67: 1–15. Peyronnin, N., and Reed, D. (eds).

Pickett, S.T.A., Cadenasso, M.L., and Grove, J.M. 2004. Resilient cities: Meaning, models, and metaphor for integrating the ecological, socio-economic, and planning realms. *Landscape and Urban Planning* 69: 369–384. doi:10.1016/j.landurbplan.2003.10.035.

Reed, M.S. 2008. Stakeholder participation for environmental management: A literature review. *Biological Conservation* 141: 2417–2431. doi:10.1016/j.biocon.2008.07.014.

Rossman, L. 2015. *Storm Water Management Model Reference Manual, Volume I—Hydrology.* U.S. Environmental Protection Agency, Cincinnati, OH.

Salvo, J., et al. 2006. New York City Population Projections by Age/Sex & Borough 2000–2030 (Briefing Booklet). New York City Department of City Planning, New York.

Sanderson, E.W. 2009. *Mannahatta: A Natural History of New York City.* New York: Abrams.

Schoemaker, P.J.H. 1995. Scenario planning: A tool for strategic thinking. *MIT Sloan Management Review* 37: 25–40.

Taylor, K.E., Stouffer, R.J., and Meehl, G.A. 2011. An overview of CMIP5 and the experiment design. *Bulletin of the American Meteorological Society* 93: 485–498.

U.S. Fish & Wildlife Service. 2011. *Science behind the Sea Level Affecting Marshes Model (SLAMM).* Washington, D.C.: U.S. Fish and Wildlife Service.

Vörösmarty, C.J., Willmott, C.J., Choudhury, B.J., Schloss, A.L., Stearn, T.K., et al. 1996. Analyzing the discharge regime of a large tropical river through remote sensing, ground-based climatic data, and modeling. *Water Resources Research* 32: 3137–3150.

Walker, B., Carpenter, S., Anderies, J., Abel, N., Cumming, G.S., et al. 2002. Conservation ecology: Resilience management in social-ecological systems: A working hypothesis for a participatory approach. *Conservation Ecology* 6: 14.

Westerink, J.J., Luettich, R.A., Jr., and Scheffner, N.R. 1993. ADCIRC: An advanced three-dimensional circulation model for shelves, coasts and estuaries, Report 3: Development of a tidal constituent data base for the Western North Atlantic and Gulf of Mexico, Dredging Research Program Technical Report DRP-92-6, U.S. Army Corps of Engineers, Vicksburg, MS.

Wheeler, D.A. 2015. Why Open Source Software/Free Software (OSS/FS, FOSS, or FLOSS)? Look at the Numbers! http://www.dwheeler.com/oss_fs_why.html. Accessed 11/28/15.

Wilson, R. 2008. The Finite Volume Coastal Ocean Model (FVCOM) applied to Jamaica Bay. Presented at the State of the Bay Symposium, New York City Department of Environmental Protection, New York, NY.

Wilson, R.E., and Flagg, C.N. In prep. Circulation of Jamaica Bay driven by buoyancy and tides simulated using a Finite Volume Coastal Ocean Model. (Prepared for submission to *Journal of Geophysical Research*).

Woelfle, M., Olliaro, P., and Todd, M.H. 2011. Open science is a research accelerator. *Nature Chemistry* 3: 745–748. doi:10.1038/nchem.1149.

9

Green Infrastructure as Climate Change Resiliency Strategy in Jamaica Bay

Maria Raquel Catalano de Sousa, Stephanie Miller,
Michael Dorsch, and Franco A. Montalto

Green infrastructure (GI) is a term used to describe a variety of natural, designed, or restored ecosystem features that are recognized to provide useful services. The first word, "green," refers to the biotic components of these systems, and "infrastructure" implies that GI must be planned, sited, shaped, and managed very carefully to address very explicit sets of urban needs, often in a decentralized manner, and sometimes in hybrid configurations with traditional "hard" infrastructure. The goal of this chapter is to explore how strategic retrofitting of specific types of GI into the Jamaica Bay watershed can potentially enhance its resilience, specifically with respect to the increasingly acute climate stressors.

Previous experience already suggests that, in appropriate configurations, GI can be, and is, used locally to reduce runoff, attenuate pollutants, control flooding, provide shading, minimize erosion, provide beauty, and become a destination for people and wildlife (See discussion below). However, the vulnerabilities exposed by recent extreme climate events and the challenges associated with recent international climate agreements have generated new interest in GI as a means of helping dynamic urban populations cope with sea level rise, rising temperatures, changing precipitation patterns, and other climate risks.

Resiliency (or resilience) is a property describing how a system reacts to a perturbation or stressor (Tschakert et al., 2013). The more resilient a system is, the larger the disturbance it can endure while maintaining its function (Hunter, 2011). Resilient systems need not necessarily be dynamic but could undergo continuous changes within an acceptable envelope of variability. The resilience of an *urban* ecosystem is measured by its ability to continue providing service levels to urban populations at acceptable levels, despite accelerated perturbations and stressors (Alberti and Marzluf, 2004) If particular infrastructure investments in an urban ecosystem contribute to an increased vulnerability of its populations to natural and human-made disasters and other drivers of change, the region's resilience would appear to be decreasing. If, by contrast, the ecosystem can be reconfigured so as to safely endure such stressors, even as they become more severe, its resilience is increasing.

A conceptual model of urban ecosystem resilience derived from more general resilience models presented in Hunter (2011), Peterson (2000), and Tschakert et al. (2013) is shown in figure 9-1. In this model, urban ecosystem resilience is illustrated by a ball in a basin. Small perturbations do not prohibit the ball from moving back to the equilibrium state (the bottom of the basin). More extreme stressors can, however, push the ball outside of the basin, causing it to eventually settle somewhere else—a disturbed state, with a new equilibrium point. This chapter discusses whether investments in GI can help specific urban ecosystems such as Jamaica Bay continue to provide services associated with their current equilibrium state (represented by the depth of the basin in figure 9-1). A related and important question is whether GI investments can actually enhance the region's resilience above current conditions.

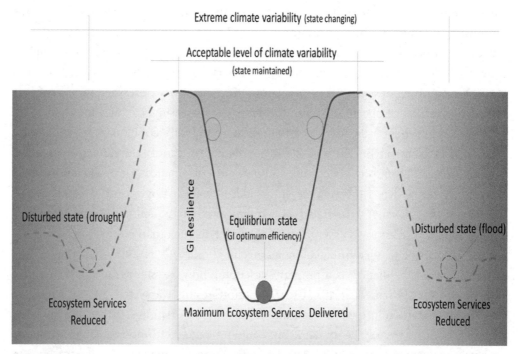

Figure 9-1. A conceptual model of urban ecosystem resilience to climate change. Courtesy of Maria Raquel Catalano de Sousa and Franco Montalto of Drexel University.

This chapter begins by reviewing the evolving meaning attributed to the term "green infrastructure," beginning with its early association with natural areas conservation, followed by later usage in the context of decentralized urban stormwater management. Next, the chapter synthesizes the climate change risks facing the Jamaica Bay region and provides illustrative examples of how GI can be used to address, mitigate, and go beyond such threats. Finally, research needs are described to increase the capacity of GI in promoting resilience in Jamaica Bay.

Evolving Definitions of Green Infrastructure

The earliest usage of the phrase "green infrastructure" in the United States was in the 1990s, with reference specifically to the conservation and restoration of what were perceived to be ecologically valuable land areas. GI was often mentioned as a means of controlling urban sprawl and its consequences, specifically landscape fragmentation and biodiversity loss (Ahern, 1995). GI often referred to interconnected networks of natural areas (e.g., open spaces, waterways, wetlands, watersheds, woodlands, wildlife habitat, parks, and greenways) believed to provide valuable ecosystem services. These services would sustain air and water resources and contribute to enriched health and quality of life for people (President's Council on Sustainable Development, 1999). Networked GI was sometimes conceptualized as interconnected hubs and links. Hubs were viewed as the most ecologically valuable areas, providing an origin and destination for wildlife and serving as a nucleus for various natural processes (Williamson, 2003). Links represented transport systems between hubs that facilitated animal and seed movement to maintain viable and persistent metapopulations (Weber and Wolf, 2000).

Though early usage of GI typically referred to networked green spaces in suburban or rural contexts, forward-thinking infrastructure and natural resource managers started to recognize the potential value of *urban* GI around the same time (table 9-1). For example, studies undertaken in the 1990s by the New York City (NYC) Department of Environmental Protection (NYCDEP) recognized for the first time that natural systems such as wetlands could reduce the cost of clean water compliance in Jamaica Bay from $2.3 billion to $1.2 billion, or nearly 50 percent. NYCDEP staff proposed wetland restoration projects around the bay's perimeter to improve water quality; they also proposed shallowing portions of the bay to reduce its residence time from thirty-five to eleven days (Appleton, 1995). These activities would restore submerged and emergent aquatic habitat, while the reduced residence time would also re-oxygenate and more frequently flush the bay with ocean water. The investment in the bay's "green" elements would improve the bay's ecology, while enhancing its ability to improve water quality, a legally mandated goal (Appleton, 1995).

Though NYCDEP's ecological restoration strategies were not formally encoded into

Table 9-1. Goals and services provided by green infrastructure systems built in New York City.

Original Motivating Goal	Description of Green Infrastructure Type	Primary Services Provided	Secondary Services Provided
Compliance with water quality improvement goals in Jamaica Bay	Perimeter coastal wetlands	Shortening of bay residence time, natural filtration	Habitat creation
Oil spill in Arthur Kill	Salt marsh restoration	Habitat restoration, prevention of erosion, phytoremediation of petroleum hydrocarbons	Public participation; attracted more restoration funds through 1996 Clean Water/ Clean Air Bond Act
Rapid urbanization on Staten Island	Enhanced freshwater wetlands, constructed stormwater capture wetlands, riparian stream habitat, and stream corridors	Flood control	Natural treatment of urban stormwater
Filtration avoidance	Catskill/Delaware Watershed Protection Program	Source water protection through strategic land acquisition and protection and sustainable forestry and farming practices	Linkages between upstream and downstream stakeholders
Reduction of combined sewer overflows	Stormwater capture Greenstreets and NYCDEP GI Program	Decentralized stormwater retention	Urban beautification, biodiversity enhancement

law until the Jamaica Bay Watershed Protection Plan (JBWPP) was completed more than a decade later, GI investments during this nascent period began elsewhere in the region, in response to other exigencies of the time. On January 1, 1990, a leaking underwater pipeline that connected Exxon plants in Linden and Bayonne, New Jersey, spewed more than 554,760 gallons (2,100 m³) of No. 2 heating oil along miles of New York and New Jersey coastlines. The event, which killed 684 birds and disrupted about 20 acres (8 ha) of *Spartina alterniflora* marsh, breeding grounds for hundreds of species, led to a settlement, including $5 million dollars that New York City used to acquire and restore wetlands and other environmentally sensitive land in the Arthur Kill (Bergen et al., 2000; Gold, 1991). Called the "genesis of local restoration efforts" (NYC, 2009), the 6 acres (2.4 ha) of *Spartina alterniflora* planted along the Arthur Kill coastline was spearheaded by the Natural Resources Group (NRG), a division of the city's Department of Parks and Recreation (NYCDPR) (Bergen et al., 2000). The goals of the new plantings were to restore the habitat that was damaged by the spill, prevent additional vegetation and peat loss through surface erosion, and break down the petroleum hydrocarbons (Bergen et al., 2000). A *New York Times* (Martin, 1994) article quoted an NRG staffer's initial observations that "oil-soaked areas that were not seeded experienced less reduction in hydrocarbons" than those that were. The early successes of the Arthur Kill salt marsh restoration effort paved the way for an additional $190 million from New York State's $1.75-billion Clean Water/ Clean Air Bond Act of 1996 to the New York/New Jersey estuary, one of the region's most significant investments in GI to date (Marc Matsil, Trust for Public Land, pers. comm.).

During this same period, the city also began recognizing the value of its urban forest, and examining other forms of GI as a potential means of reducing flooding and protecting its drinking water supply. In 1991, the city received a five-year, $6.2-million-dollar grant from the Lila Wallace-Reader's Digest Fund to "preserve, protect and enhance the city-owned forests of New York City and to increase public awareness and appreciation of them." The notion that enhanced green spaces could become an alternative to hard infrastructure, however, gained more attention in Staten Island and in the city's upstate watersheds.

With the completion of the Verrazano-Narrows Bridge in 1964, large portions of Staten Island had become much more accessible for development. By the 1980s, the rapid pace of urbanization had exceeded the city's ability to build sanitary and stormwater infrastructure in the borough and drastically modified historical streams and wetlands. New development was forced to rely on septic systems for wastewater disposal, while the lack of adequate drainage, either natural or engineered, soon made flooding a pervasive problem. In 1990, after a complex public debate, the NYCDEP began construction of the Bluebelt, an integrated network of what is today about 400 acres (162 ha) of enhanced freshwater wetlands, constructed stormwater capture wetlands, riparian stream habitat, and nearly 11 miles (18 km) of stream corridors to which runoff from developed areas is diverted. Though its critics point out that the Bluebelt is fundamentally an end-of-pipe treatment solution that is a disincentive to stormwater source control measures, this natural treatment system conveys runoff away from homes and businesses, averting flood damage—worth $80 million to the city in terms of avoided storm sewer infrastructure (NYC, 2007).

In parallel, the city began to look to GI as a cost-effective way to sustain the quality of its drinking water. With passage of the 1986 Safe Drinking Water Act, large cities such as New York City were required to filter their municipal water supplies. In what became one of the most innovative international models of watershed protection, NYCDEP successfully argued that strategic land stewardship and incentive programs in its watershed would justify filtration avoidance. By acquiring and protecting land, and incentivizing sustainable forestry and farming practices, the city could "prevent contaminants from reaching water sources" (NYCDEP, 2016). It is estimated that the Catskill/Delaware Watershed Protection Program has saved the city between $4 billion and $8 billion (Alcott et al., 2013) in avoided infrastructure costs, and New York City remains today one of only five major cities that has avoided nonnatural filtration interventions (NYC, 2007).

Despite these early examples, GI did not garner more widespread urban use until 2006, when the Natural Resources Defense Council (NRDC) published a seminal report defining GI as "trees, vegetation, wetlands, and open space preserved or created *in developed and urban areas* [emphasis added]—a strategy for stopping water pollution at its source" (NRDC, 2006). The focus of the NRDC definition was on how GI could be used for urban stormwater source control, as part of a new decentralized strategy for avoiding the construction of large in-line storage tanks, tunnels, and end-of-pipe treatment systems at separate and combined sewer overflows (CSOs).

Following publication of the NRDC report, stormwater utilities and other government agencies around the nation began seriously considering GI as a means of reducing runoff through small-scale rain gardens, green roofs, bioswales, constructed wetlands, and permeable pavements, among other strategies. Early studies involved both modeling and implementation of a variety of pilot projects. One of the earliest studies, funded by the National Oceanic and Atmospheric Administration, demonstrated how curbside planter beds could be used to divert stormwater away from combined sewer catch basins along a six-block hypothetical "Green Corridor" in the Bronx (The Gaia Institute, 2016). Though this project was never built, the city was about to undergo a dramatic paradigm shift in how it managed its stormwater.

Jump-starting this shift were natural resource managers at NYCDPR, who began to divert street and sidewalk runoff to small pocket parks known as Greenstreets. Though urban beautification was the original goal of the Greenstreets program, NYCDPR recognized that the soil and vegetation inside the Greenstreets could be used for infiltration and evapotranspiration of stormwater, reducing the load that urban runoff presented on the city's combined sewer system. The first stormwater capture Greenstreet was built on the southeast corner of 110th Street and Amsterdam Avenue in Manhattan in 2006. In the Jamaica Bay watershed, the first stormwater GI systems were Greenstreets built in 2010, with funding earmarked by the American Recovery and Reinvestment Act for "shovel-ready" infrastructure (figure 9-2).

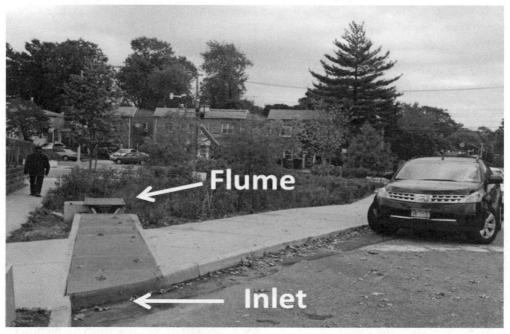

Figure 9-2. The inlet of the first stormwater-capture Greenstreet built by the City of New York's Department of Parks and Recreation in the Jamaica Bay watershed. Runoff generated within a tributary area that is 3.8 times the size of the vegetated area enters the site through the curb-cut inlet (indicated). Courtesy of Maria Raquel Catalano de Sousa and Franco Montalto of Drexel University.

The importance of such initiatives for restoring Jamaica Bay had already been identified in the Jamaica Bay Watershed Protection Plan, completed in October 2007 (NYCDEP, 2007). Along with the ecological restoration of the bay's marsh islands, beaches, and dunes, and oyster, eel grass, and ribbed mussel habitat, the JBWPP identified "stormwater source controls" as a potentially effective way of reducing CSOs and other untreated discharges to the bay. Modeling presented in the JBWPP suggested that rain barrels, rain gardens, and swales could reduce untreated discharges to the bay by 6 percent and that GI implemented on large rooftops (>4,950 square feet [460 m²]) could reduce discharges by approximately 13 percent; however, a more aggressive GI strategy targeting new and existing development could reduce bay discharges by as much as 24 percent (NYCDEP, 2007). The plan recommended that monitored pilot projects be implemented on roadways, streets, sidewalks, and vacant lands, as well as in parks, tree pits, and in more Greenstreets.

Building off of these and other early successful demonstrations and studies, New York City, like many other municipalities, began to formally encode GI into its stormwater and CSO control plans. Citing the effectiveness of stormwater capture with Greenstreets and the Bluebelt, PlaNYC established as a goal the expansion, tracking, and analysis of "new

Best Management Practices on a broad scale" in 2007, calling for an interagency task force to integrate GI planning citywide. In 2008, the Mayor's Office of Long-Term Planning and Sustainability published the city's Sustainable Stormwater Management Plan, which formally recognized that "stormwater runoff does not have to be an inevitable by-product of development" and that "building and landscape designs that mimic natural systems, and infiltrate, retain, or detain rainfall on site can reduce excess flows into our sewers, streets, and waterways." In September 2010, the NYCDEP released its Green Infrastructure Plan, which committed the city, by 2030, to capturing the first inch (25 mm) of rainfall from 10 percent of the impervious areas served by combined sewers with GI systems. It estimated that GI can reduce CSOs by approximately 1.5 billion gallons (5.7 million m³) per year at a cost of approximately $2.4 billion (NYCDEP, 2014).

Using GI to Increase Resilience and Reduce Climate Risks

Though the GI projects described above undoubtedly help the city to respond incrementally to climate change, most of these programs were not conceived expressly for this purpose. The earliest reference to GI as a potential climate change adaptation strategy was in 1999 when the President's Council on Sustainable Development recommended it as one of several climate protection strategies offering multiple benefits that help to solve social, economic, and environmental problems while creating opportunities for current and future generations. Carbon sequestration, recreation, and flood and erosion prevention and control, for example, were cited as key benefits of green space preservation and urban forestry (President's Council on Sustainable Development, 1999).

In New York City, reference to GI as a potential climate change adaptation strategy began in earnest in the months following Hurricane Sandy. As reports documenting the protective value of beaches, dunes, and wetlands in places such as Nags Head, North Carolina (Mignoni, 2014), and Westhampton Beach (Bocamazo et al., 2011), Point Lookout, Lido Beach, and Atlantic Beach, New York (Navarro and Nuwer, 2012), during Sandy and previous coastal storms surfaced, the planning and design community began to give a critical look to GI systems as an alternative to hard forms of coastal protection, such as breakwaters and dikes.

In June 2013, the U.S. Department of Housing and Urban Development launched the Rebuild by Design competition (RBD, 2016), which for the first time focused the attention of designers, researchers, community members, and government officials on physical, social, and ecological dimensions of resiliency. Many of the winning proposals included various forms of GI. The Hunts Point Lifelines team, for example, designed GI systems for the Hunts Point Market that could detain upstream runoff generated during 100-year precipitation events, even if accompanied by high tides and surges such as those experienced during Sandy. The Living Breakwaters team proposed a "necklace" of living breakwaters, underwater GI systems that would protect the city's coasts against wave damage, flooding, and erosion, with programming linking these systems to the local community. These and other similarly focused proposals helped infrastructure managers to envision how GI systems could become important components of regional resilience plans, from both ecological and social perspectives.

Various government agencies followed suit. Since Hurricane Sandy, the U.S. Army Corps of Engineers (USACE) has initiated significant dune and beach restoration projects along the large stretches of New York City coast (Gardner, 2013). On Rockaway Beach, the USACE effort is replacing roughly 95 million cubic feet (2.7 million m³) of sand to reduce risks from future storms (Gardner, 2013). This volume includes sand lost during Hurricane Sandy, as well as sand lost to wind and wave erosion since the last renourishment project during 2004 (Gardner, 2013). In addition to beach replacement, the USACE will build and maintain a dune system around New York City, with the system being continuous along the Rockaway Peninsula.

After releasing Vision 2020, a blueprint for how New York City's urban waterfront could be redesigned to promote urban sustainability and resilience in 2011 (NYCDCP, 2011), the Department of City Planning (NYCDCP) issued its Urban Waterfront Adaptive Strategies in 2013, which specifically addressed the threats of climate change and sea level rise (NYCDCP, 2013). That same year, the city's Special Initiative on Recovery and Resilience released A Stronger, More Resilient New York, which committed the city to rebuilding and restructuring the coast to be more resilient to the threats of storm surge, wave damage, and coastal flooding during both routine and extreme events (NYC, 2013). The city's total commitment to the waterfront amounted to more than $20 billion and called for the use of green, gray, and hybrid infrastructure (NYC, 2013). GI mentioned in the plan included primarily beach nourishment, dune construction and stabilization, and the creation and maintenance of wetlands (NYC, 2013; NYCDCP, 2013). Additional efforts call for the restoration and creation of living shorelines, oyster beds, and marsh islands (TNC, 2015; Schuster and Doerr, 2015). The city committed $12 million to restore a city-owned wetland in Staten Island (Office of the Mayor, 2014) and the U.S. Environmental Protection Agency (U.S. EPA) has provided grants to NYCDPR to protect, restore, and monitor salt marshes, including new designs for Jamaica Bay (U.S. EPA, 2014).

Though the traditional services that infrastructure planners expected from GI (e.g., water quality improvement, flood and erosion control, habitat restoration) are still needed, the plans and policies described above suggest increased attention to GI as a means to promoting resilience. Hurricane Sandy jump-started a wide-ranging policy discussion regarding how GI might be able to help make the region more resilient to *coastal* climate risks, but the region faces a range of climate risks beyond the coast. In addition to sea level rise, the region is facing rising temperatures and changes in the timing and amount of precipitation. For GI to promote comprehensive resilience, it would ideally help the region adapt to a variety of climate futures. Adaptation, here, refers both to the ability of the region's infrastructure systems, people, and wildlife to physically cope with a changing climate, and to the behavioral response of its residents, institutions, and governance structures to the challenges that climate and other stressors present. The following section reviews trends in sea level rise, temperature, and precipitation facing Jamaica Bay, highlighting the challenges and opportunities that each presents with regard to GI.

GI for Resilience to Sea Level Rise, Surges, and Waves

Over the past thousand years, sea level along the Atlantic Seaboard has risen at a rate of 0.34–0.43 inch (0.86–1.09 cm) per decade; during the twentieth century this rate increased to 1.2 inch (3.05 cm) per decade (NCA, 2013). The rate of rise along the U.S. Atlantic coast was, in general, greater than the global average during the last century (Yin et al., 2011; Sallenger et al., 2012) and is projected to remain so over the twenty-first century (NCA, 2013). Locally, however, the rate of sea level rise varies due to uneven rates of marine sediment deposition, glacio-isostatic adjustment (Mitsch and Gosselink, 2000; Montalto and Steenhuis, 2004), and land subsidence (Church et al., 2010), as well as changes in ocean circulation in the North Atlantic (Sallenger et al., 2012). The New York City Panel on Climate Change (NPCC, 2013) reports that by the 2050s, the thirty-year mean sea level in the city could be 0.92–2.6 feet (0.28–0.78 m) higher than it was during a baseline observation period (2000–2004). Rising sea level increases risks posed by waves, surge, and periodic extreme events. As this trend progresses, urban coastal flooding is expected to worsen, causing disruptions to services and threatening public health and safety (NCA, 2013).

Beaches, Dunes, and Wetlands

GI systems that can be used to address sea level rise include beaches, dunes, and wetlands. As sea levels rise, beach nourishment will become particularly important as sedimentation will likely not be able to keep up with the rate of sea level rise, particularly in systems such as Jamaica Bay, where the natural sedimentation process has been significantly reduced due to consistent dredging. One disadvantage is that many beaches will need to be renourished on a regular basis, as the kinetic energy they dissipate from the waves promotes sand erosion. Despite their need for ongoing maintenance, enhanced beaches also carry significant recreational value to the city's residents and economic value for the regional economy. The work involved in building and maintaining them has the potential to create local jobs.

Dunes proved their value as a coastal protection mechanism during Hurricane Sandy, especially where they were large and continuous. Their absence was also particularly apparent, for example in Long Beach, New York—the community opposed the construction of a $7-million-dollar dune project only to suffer $200 million in damages during Sandy (Navarro and Nuwer, 2012). Dunes can also be effective against the more frequent, nonextreme coastal storms that bring small surges and damaging waves that the city experiences multiple times a year. Like beaches, dunes also require regular maintenance to replace sand eroded by wind and waves, though erosion processes can be minimized by establishing vegetation along the top and back side of dunes to help maintain the integrity of the system. This same vegetation also has significant habitat value, including for endangered species such as seabeach amaranth (*Amaranthus pumilus*).

Coastal wetlands have been estimated to provide more than $23 billion per year in storm protection services in the United States by reducing surges, attenuating waves, and retaining water to reduce flooding (Costanza et al., 2008; Gedan et al., 2011; Spalding et al., 2014; Barbier, 2015). However, scientific studies suggest that the ability of wetlands to provide coastal protection is nuanced and specific to both site and storm. During individual events, the storm protection services of wetlands appear variable and highly dependent upon wind speed, storm forcing, elevation, the surrounding coastal landscape, waterbody connectivity, and wetland type (Barbier et al., 2008; Resio and Westerink, 2008; Loder et al., 2009; Ebersole et al., 2010; Wamsley et al., 2010; Gedan et al., 2011; Acreman and Holden, 2013; Barbier and Enchelmeyer, 2014; Spalding et al., 2014). Some research even suggests that during slow-moving storms with high winds, surges and associated damages can actually be higher over wetlands than in surrounding areas (Resio and Westerink, 2008; Wamsley et al., 2010; Hu et al., 2015). This situation can arise when the forces pushing the water toward the land (a combination of wind and water pressure) significantly exceed the frictional forces imposed by the wetland surface. In such cases, the slow-moving surge can progress inland with very limited resistance.

The vegetation found in wetlands plays an important role, as increasing stem height and density both increase the ability of a wetland to provide protection from storm surges and waves (Loder et al., 2009; Spalding et al., 2014; Hu et al., 2015). This observation sets up complex trade-offs in places such as Jamaica Bay, where tall invasive *Phragmites australis* may be simultaneously providing coastal protection and other services, while also encroaching on and replacing historical stands of *Spartina alterniflora* and reducing regional biodiversity.

Wetland location is also key, as upland sites are more valuable in averting flooding associated with heavy precipitation, while lowland wetlands are more valuable as a coastal protection measure (Acreman and Holden, 2013). However, the ability of low-lying coastal wetlands to provide coastal protection may be lessened as sea level rises. Beginning about 9,000 years ago, in the last major period of sea level rise, New York City wetlands started to form as fine-grained marine sediments were deposited in drowned coastal stream and river valleys, a process known as marine transgression (Warren, 1997; Montalto and Steenhuis, 2004). Wetlands migrated farther inland as sea level got higher. Many of today's urban coasts are, however, lined with buildings, roads, and other infrastructure, presenting physical barriers to the migration of coastal wetlands—natural systems that are already sediment-starved. The result is a gradual drowning of these pulsing systems, which are inundated for longer and longer periods with each tidal cycle, becoming first mudflats and, finally, open water.

The most valuable near-term role of wetlands may be in providing protection from storm surges associated with increasingly frequent smaller high water events and protecting coastal infrastructure and ecosystems from waves, because even small, fragmented, urban wetlands are capable of providing these services. Research suggests that the degree of wave attenuation is primarily determined by wetland continuity and surface roughness, not overall wetland area or distance traversed by the wave (Barbier et al., 2008; Loder et al., 2009; Gedan et al., 2011; Barbier and Enchelmeyer, 2014).

GI for Resilience to Rising Temperatures and Heat Waves

Since 1970 mean temperature across the U.S. Northeast has increased at a rate of 32.5°F (0.25°C) per decade (Hayhoe et al., 2006). Over the next several decades, temperatures across the region (compared with 1961–1990) are expected to continue to increase by 41–44.1°F (5–6.7°C) in winter, and by 35–46°F (1.67–7.80°C) in summer, under both the higher (A1F1) and lower (B1) Intergovernmental Panel on Climate Change emissions scenarios, respectively (Frumhoff et al., 2007). By 2080, mean annual temperatures (relative to 1970–1999) are projected to increase by 34.9–41°F (1.6–5°C) (CCRUN, 2015). By 2041–2070, the number of days with a maximum temperature greater than 95°F (35°C) is projected to rise by up to fifteen days compared with 1971–2000, assuming continued increases in global emissions (NCA, 2013). New York City is specifically expected to experience more than twenty-one additional days (compared with 1971–2000) per year above 86°F (30°C) (NPCC, 2013).

Shading and Evapotranspiration by Vegetation

Vegetated GI can help cool urban environments by providing shading and enhancing evapotranspiration. During the summer, leaves and branches can diminish the amount of solar radiation that penetrates a tree's canopy by 10 to 30 percent (U.S. EPA, 2013), reducing surface warming. Through the process of transpiration, vegetation extracts water from the soil, releasing it as water vapor through its leaves. Because the conversion of water from liquid to gas requires roughly 2,000 BTU for each kilogram of liquid water (Campbell and Norman, 1998), when plants transpire they wick heat away from the surface, incrementally lowering the ambient temperature. On hot days, a tree can transpire up to 100 gallons (379 l) of water (Akbari, 2002), which is significant because the temperature of city centers can be up to 41°F (5°C) higher than surrounding leafy green suburbs, a phenomenon known as the urban heat island effect (Akbari et al., 1992). Reductions of up to 66.2°F (19°C) have been observed in a study comparing surface temperature of asphalt and tree-shaded areas of an urban park (Rahman et al., 2014).

An interesting co-benefit arises when vegetated GI is used to manage stormwater. Preliminary research by the authors suggests that Greenstreets and other engineered GI systems evapotranspire more water when they are connected to larger tributary drainage areas. This phenomenon is likely because actual evapotranspiration rates in the Jamaica Bay region are typically moisture- and not energy-limited. That is, the region's microclimate would enable more evapotranspiration to occur if there were more root zone soil moisture available to plants. Each time it rains, vegetated green spaces that are hydraulically linked to off-site impervious tributary drainage areas get wetter than hydraulically isolated green spaces (e.g., separated by a curb, as pre-2006 Greenstreets were). This linked configuration leads to higher relative soil moisture levels, allowing plants found there to evapotranspire at higher rates than vegetation found in standalone sites, reducing the local heat island *while* managing stormwater. As a corollary, within a given GI site, more evapotranspiration occurs at the lowest elevation of vegetated GI systems, because water traveling over the surface tends to pond and infiltrate more frequently and to a greater extent. Other work suggests significant inter- and intraspecies differences in the transpiration rates, for individual plants subjected to both the same soil moisture and climatic conditions (Miller, 2014).

These observations hint at other interesting multifunctional water redirection possibilities not only in Greenstreets but also on green roofs, green walls, and other vegetated and nonvegetated surfaces. A pilot project in Paris, France, actually applies water onto paved street surfaces during hot summer days specifically to trigger evaporation and the associated cooling. Rooftop misters and sprinklers, increasingly found in semiarid cities, can have the same effect at the building scale. The opportunity for multiple services arises to the extent that the recycled water consists of locally harvested stormwater and not potable water.

Vegetation can also help reduce the use of air conditioning in buildings. In a study conducted in New York City neighborhoods, a combined strategy of tree planting and green roofs was associated with peak load reductions of 2–3 percent (Gaffin et al., 2012). Saiz and colleagues (2006) compared energy use under a green versus conventional roof and found that although annual energy savings are just more than 1 percent, the summer cooling load is reduced by more than 6 percent, with associated reductions of up to 25 percent in the peak-hour cooling load of the upper floors. A study performed in a Pittsburgh, Pennsylvania, commercial building revealed that during summer months, surface temperatures on a gravel ballasted roof membrane can be up to 50 percent higher than on those of a green roof (Kosareo and Reis, 2007).

For all of these benefits to be significant, extensive canopy coverage needs to be present in urban neighborhoods. The JBWPP reported that the Jamaica Bay watershed consists of approximately 65–70 percent impervious surfaces, with the 26th Ward sewershed as much as 83 percent impervious. In such ultraurban contexts, space constraints limit the number of new trees, yards, and parks that can be built, and engineered GI systems such as green roofs and green walls become of greater local interest.

Of concern to some GI planners are the risks that elevated temperatures, prolonged drought, and related phenomena create for new urban vegetation, because plant mortality triggers the need for plant replacement, increasing maintenance costs. Initial studies with a limited number of species selected from NYCDPR's GI plant pallet (Catalano de Sousa, 2015) suggest that through judicious plant selection, the risk of plant mortality can be kept to a minimum. Additional studies are required focusing on the full extent of new vegetation being introduced to the urban environment. Research efforts also need to scale up the energy benefits of evapotranspiration computing, for example, the regional heat island mitigation values for different incremental increases in percent vegetative cover.

GI for Resilience to Changes in Precipitation Patterns

Climate models predict that future warming will have a more pronounced impact on storm intensity than on total annual precipitation depths, though both are expected to increase. Using 1961–1990 as a reference period, the northeast United States is projected to witness by 2100 an increase of 10–15 percent in precipitation intensity (the amount of rain that falls on any given day), 12–13 percent in the number of heavy precipitation events (defined as more than 2 inches [5.08 cm] of rain in 48 hours), 20 percent in the intensity of once-a-year extreme precipitation events (Frumhoff et al., 2007), and up to 14 percent in annual precipitation amounts (Hayhoe et al., 2006). Another assessment (NCA, 2013) suggested that the Northeast has experienced a greater increase in extreme precipitation over the past five decades than any other region in the country. Between 1958 and 2011, this region experienced a 74 percent increase in the amount of precipitation falling in very heavy precipitation events (defined as the heaviest 1 percent of all daily events).

By the end of this century and under a high emissions scenario, winter precipitation in the region is projected to increase from 5 to 20 percent (NCA, 2014), while summer precipitation is actually expected to decrease by up to 2 percent (Hayhoe et al., 2006). Less frequent but more intense precipitation events may increase runoff and exacerbate the severity of drought, simultaneously reducing evapotranspiration and its associated climate-regulating benefits. Shifts in the timing of precipitation, accompanied by simultaneous warming trends, can reduce winter snowpack, leading to droughts in the spring and summer. This same trend is likely to increase the risk of winter flooding, especially in impervious urban areas. In summary, though total annual precipitation amounts may not change significantly, there could be significant increases in the occurrence of floods and droughts throughout the region (NCA, 2013; Trenberth et al., 2003).

Stormwater GI

By promoting interception, infiltration, and evapotranspiration, rain gardens, green roofs, bioswales, stormwater wetlands, and other stormwater GI can help to mitigate the effects of altered precipitation characteristics. Under recent precipitation conditions, the NYCDEP (2012) reports that GI pilot projects (bioretention and enhanced tree pits), monitored between 2011 and 2012, retained between 64 percent and 100 percent of the stormwater routed to them during storms of up to 1 inch (25.4 mm). The same report featured a pair of connected bioretention areas that was able to retain 100 percent of the stormwater routed to them during events of 2 inches (50 mm) and less, while retaining up to 80 percent of stormwater inflows generated by 4–8 inch (100–200 mm) events.

GI monitoring by the authors suggests similar stormwater capture levels for a wide range of GI systems. The 2,874-square-foot (267 m²) lined ABC Carpet constructed wetland in the Bronx, which receives runoff from a parking lot eleven times its size, completely retained runoff from all but eight of sixty-one monitored events over two years, making the median percent retention rate of the facility 100 percent. Monitoring of the 398-square-foot (37 m²) Shoelace Park rain garden confirmed that this Bronx GI site retained more than 80 percent of the total runoff entering it during most storms from a tributary area more than fifteen times its size (Feldman, 2015). Preliminary analysis of the Poppenhusen rain gardens in College Point, Queens, suggests 100 percent retention of all events monitored over a two-year period. Monitoring apparatus installed on the Jacob K. Javits Convention Center in Manhattan indicates that approximately 72 percent of the precipitation that fell on this 1-inch-thick (2.5 cm) green roof system during monitoring was captured and retained.

Regional climatic projections, however, forecast larger and more intense precipitation events in the future. GI monitoring results across all sites mentioned above suggest that the percent of runoff retained in a given GI facility is reduced for larger storms. However, research conducted at the Nashville Boulevard Greenstreet located in the Jamaica Bay watershed suggests only a modest reduction in performance during extreme events (figure 9-3). Because of inlet bypass, the site retained an average of only 70 percent of the runoff generated in its tributary catchment area during each of ninety-two different events over a four-year monitoring period, with some evidence that more frequent maintenance visits increased stormwater capture levels. All of the runoff that entered this Greenstreet during any of the storms was infiltrated, with the exception of a short ten-minute period

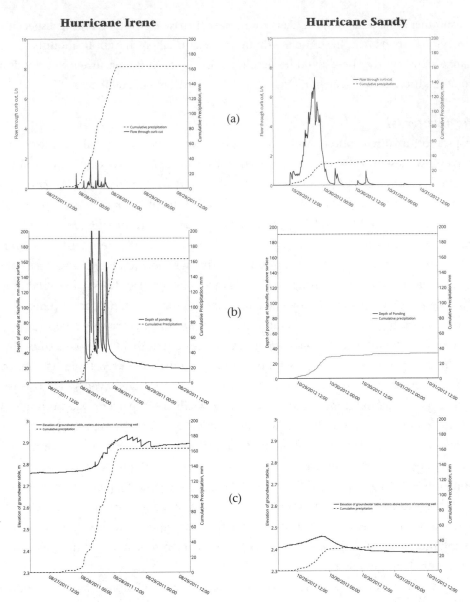

Figure 9-3. Hydrologic performance of the Nashville Boulevard Greenstreet during two recent extreme events, Hurricane Irene in August, 2011 (left), and Hurricane Sandy in October, 2012 (right). In all six charts, the cumulative precipitation during the storms is depicted on the right axis; in total 163 mm (6.4 inches) of rainfall during Irene, and 33 mm (1.3 inches) during Sandy. The upper charts (a) depict inflow to the Greenstreet as measured at the inlet, during Irene (left) and Sandy (right). The middle charts (b) depict the depth of water ponded within the site during each of the two storms. The horizontal line above the 180 mm line represents the depth of ponding required for system overflow. The lower charts (c) depict slight changes in the water table elevation below the Greenstreet as a result of infiltration. Courtesy of Maria Raquel Catalano de Sousa and Franco Montalto of Drexel University.

during Hurricane Irene when the site briefly overflowed. When the events were organized by depth and intensity, only a small reduction in performance levels was observed. The percent of tributary runoff captured was reduced from an average of 77 percent of total runoff for nonextreme events, to 60 percent for extreme events. Though this site receives runoff from a tributary area that is only four times its size and is underlain by high permeability sandy soils, these results suggest that engineered stormwater GI systems may help the region reduce urban flooding during extreme precipitation conditions. Additional research is needed to test the extreme event performance of other GI systems installed in other hydraulic loading and soil conditions (Catalano de Sousa, 2015).

As more small-scale GI systems are implemented throughout the region, maintenance is becoming a significant challenge. Well-maintained GI systems can more reliably regulate stormwater flows, and though maintenance of a distributed network of small GI systems can be challenging for organizations such as the NYCDEP, this task may represent an opportunity for engaging and potentially employing the public in environmental management. There is increasing interest in the opportunity that distributed infrastructure represents for "green-collar" jobs training and for generating a new cohort of community-based green jobs (Montalto et al., 2012).

Concluding Thoughts

The experiences and research summarized in this chapter suggest that different forms of GI can help urban communities such as those of Jamaica Bay to control floods, improve water quality, manage stormwater, reduce CSOs, protect drinking water supplies, and enhance habitat, while also creating new recreational opportunities and potentially even jobs and job-training opportunities. The same GI systems also seem to carry great potential for helping the region to adapt to dynamic and nonstationary climatic conditions (table 9-2).

Although there are many infrastructure strategies that can provide each of these services individually, GI systems are unique in their ability to provide multiple services at the same location. In this *multifunctionality*, GI is an intrinsically *cost-effective* component of the urban infrastructure network. Multifunctionality is one of five urban planning and design strategies listed by Ahern (2011) as crucial for building urban resilience.

GI systems can also help address four other strategic goals identified by Ahern (2011): modularity, adaptiveness, diversity, and multiscale connectivity. Because GI systems are distributed throughout the landscape, they must be *modular*, meaning they are subdivided into independent elements that can be deployed singularly or together. Lessons learned at one site can be used to improve the chances of success at another site, and in so doing help to gradually *adapt* designs to changing local conditions. Because they are living systems, GI systems have the potential to be *biologically diverse*. As distributed infrastructure, their application must also be *spatially diverse*. Their modularity, adaptability,

Table 9-2. Green infrastructure strategies that may enhance resilience to different climate stressors.

Climate Stressor	Associated Green Infrastructure Strategy	Resilience Services Provided
Sea level rise	Beaches, dunes, wetlands	Surge and wave buffering
Increased temperature	Vegetated patches, especially those connected to off-site tributary drainage areas	Shading, evapotranspiration
Changed precipitation patterns	Greenstreets, bioswales, green roofs, constructed wetlands, rain gardens, permeable pavements	Stormwater capture, flood control

and diversity all help to create *redundancy,* i.e., one GI facility, built at a particular time in one part of the city, can fail without significantly hindering the ability of the entire distributed network to continue to function. GI systems are also deliberately positioned at the interfaces that exist between the land and the water, the surface and the subsurface, and the built and natural components of urban ecosystem. They are appreciated by multiple stakeholders (both infrastructure managers and residents) for both their use and nonuse values. The services they provide are valuable locally, but also, in an incremental way, at both regional and global scales. As such, they represent crucial *connective, multiscale networks* that can be conserved, engineered, or enhanced within urban environments, building simultaneously both ecological and social resilience.

Although these attributes of GI systems are encouraging, the systems need to be widely implemented in Jamaica Bay to fundamentally change the conditions that determine the region's resilience. Widespread implementation of GI remains challenging due to a number of formidable, persistent, and common, though not insurmountable, obstacles (Montalto et al., 2012). Combined sewers continue to discharge untreated waste into the bay, compromising water quality. To reduce overflows, decentralized stormwater GI would need to be integrated into already developed neighborhoods, where imperviousness is actually increasing with expanded development of buildings, roads, and other infrastructure in floodplains. To improve bay water quality, wetlands need to be restored along hardened shorelines that offer limited inland migration space. Deep channel dredging operations continue to disrupt natural sedimentation patterns, reducing the ability of existing GI to accrete in proportion to sea level rise (ARCADIS, 2014; NYCDEP, 2014). Any form of new bird habitat creates safety concerns for air traffic coming into and out of John F. Kennedy International Airport, one of the busiest airports in the world (Sewell, 2011; ARCADIS, 2014; Tennant, 2014).

Tackling such issues requires broad-based institutional commitment and will, flexibility, and, of course, significant financial resources. Fortunately, there is reason to believe that these resources are all growing. More than $100 million has already been allotted for coastal GI within Jamaica Bay, with efforts spread out over many organizations (NYCDCP, 2013; NYCDEP, 2014; TNC 2015). Hybrid strategies have been proposed at Howard Beach, including restoration of marshland both inland and on marsh islands, as well as the use of ribbed mussel beds to dampen waves and provide water quality benefits (NYCDEP, 2014; TNC, 2015). The National Park Service and NYCDPR are working together to maintain public access to nature, recreation, and education opportunities for both local visitors and tourists (NYCDEP, 2014; Siegman, 2015). This includes support of the Jamaica Bay-Rockaway Parks Conservancy and their commitment to restoring 10,000 acres (4,407 ha) of public parkland (NYCDEP, 2014). NYCDEP and NYCDPR are also looking to use GI in Jamaica Bay for stormwater capture, water quality improvements, and wave attenuation. They are planning to build ribbed mussel beds and floating wetlands at Fresh Creek and Brant Point, respectively (NYCDEP, 2014). Plans for more traditional gray infrastructure (such as bulkheads and sea walls) can potentially be modified using ecofriendly materials to double as habitat for benthic invertebrates (ARCADIS, 2014). Other significant obstacles warranting focused attention involve coordination between the various stakeholders and a suite of complex regulatory and legal hurdles.

For GI to be instrumental in maintaining, enhancing, and growing the functions and services it is already providing, more work is necessary at institutional, technical, and community levels. Institutionally, forward thinking and adaptive plans, policies, and incentives are needed to test new strategies that may help to overcome obstacles. Technically, more work is necessary documenting the performance and efficiency of multifunctional GI strategies and their hybrid cousins. At the community level, greater participation of the public in decision making, and even in the maintenance and monitoring of engineered GI systems, is necessary. Urban landscapes are both habitat for wildlife and homes for people. Any effort to make them more resilient must emerge from a deep understanding of local needs, capacities, desires, and changes.

Finally, although this chapter documents various ways that GI can contribute to regional resilience, GI systems can themselves be vulnerable to climate change. We need to develop a more comprehensive understanding of how climate change will affect the specific species commonly used in GI facilities and also to study how these systems function under a wide range of interacting and complex climatic and anthropogenic stressors that are already evident in the dynamic urban environment that surrounds today's Jamaica Bay.

Note: Portions of this chapter have been adapted from the doctoral dissertation of Maria Raquel Catalano de Sousa and from a research report investigating the relationship between GI and building damages during Hurricane Sandy, prepared by Miller and Montalto for the Trust for Public Land. The authors thank Bram Gunther and Marit Larson, from New York City Department of Parks and Recreation, and Eric Rothstein, from eDesign Dynamics, for their comments on earlier versions of this chapter.

References

Acreman, M., and Holden, J. 2013. How wetlands affect floods. *Wetlands* 33: 773–786. doi:10.1007/s13157-013-0473-2.

Ahern, J. 1995. Greenways as a planning strategy. *Landscape and Urban Planning* 33: 131–155.

Ahern, J. 2011. From fail-safe to safe-to-fail: Sustainability and resilience in the new urban world. *Landscape and Urban Planning* 100(4): 341–343.

Akbari, H., Davis, S., Dorsano, S., Huang, J., and Winnett, S. (eds). 1992. *Cooling Our Communities: A Guidebook on Tree Planting and Light-Colored Surfacing.* Washington, DC: U.S. Environmental Protection Agency, 26.

Akbari, H. 2002. Shade trees reduce building energy use and CO_2 emissions from power plants. *Environmental Pollution* 116: S119–S126.

Alberti, M., and Marzluf, J.M. 2004. Ecological resilience in urban ecosystems: Linking urban patterns to human and ecological functions. *Urban Ecosystems* 7(3): 241–265. doi:http://dx.doi.org/10.1023/B:UECO.0000044038.90173.c6.

Alcott, E., Ashton, M.S., and Gentry, B.S. 2013. *Natural and Engineered Solutions for Drinking Water Supplies: Lessons from the Northeastern United States and Directions for Global Watershed Management.* Boca Raton, FL: Taylor & Francis.

Appleton, A. 1995. The Challenge of Providing Future Infrastructure in an Environment of Limited Resources, New Technologies, and Changing Social Paradigms. Proceedings of a Colloquium. *Horizontal Integration of Infrastructure Services: The New York City Experience*, 94.

ARCADIS. 2014. Coastal Green Infrastructure Plan for New York City. Report for the Hudson River Estuary Program, New York State Department of Environmental Conservation, with support from the New York State Environmental Protection Fund, in cooperation with the New England Interstate Water Pollution Control Commission. Available at: http://www.dec.ny.gov/docs/remediation_hudson_pdf/cginyc.pdf.

Barbier, E.B. 2015. Valuing the storm protection service of estuarine and coastal ecosystems. *Ecosystem Services* 11: 32–38.

Barbier, E.B., and Enchelmeyer, B.S. 2014. Valuing the storm surge protection service of US Gulf Coast wetlands. *Journal of Environmental Economics and Policy* 3(2): 167–185. doi:10.1080/21606544.2013.876370.

Barbier, E.B., Koch, E.W., Silliman, B.R., Hacker, S.D., Wolanski, E., et al. 2008. Coastal ecosystem-based management with nonlinear ecological functions and values. *Science* 319: 321–323.

Bergen, A., Alderson, C., Bergfors, R., Aquila, C., and Matsil, M.A. 2000. Restoration of *Spartina alterniflora* salt marsh following a fuel oil spill, New York City, NY. *Wetlands Ecology and Management* 8(2): 185–195. doi:10.1023/A:1008496519697.

Bocamazo, L.M., Grosskopf, W.G., and Buonuiato, F.S. 2011. Beach nourishment, shoreline change, and dune growth at Westhampton Beach, New York, 1996–2009. *Journal of Coastal Research* Special Issue: 59: 181–191. doi:10.2112/si59-019.1

Campbell, G.S., and Norman, J.M. 1998. *An Introduction to Environmental Biophysics*, 2nd ed. New York: Springer.

Catalano de Sousa., R. 2015. Assessing Green Infrastructure as an Effective Strategy to Help Cities to Build Resilience to Climate Change. (Ph.D. disseration, Drexel University, 2015).

CCRUN (Consortium for Climate Risk in the Urban Northeast). 2015. Climate Change Projection Maps. Accessed at: http://ccrun.org/node/59.

Church, J.A., Woodworth, P.L., Aarup, T., and Wilson, W.S. 2010. *Understanding Sea-Level Rise and Variability*. New York: John Wiley & Sons.

Costanza, R., Perez-Maqueo, O., Martinez, M.L., Sutton, P., Anderson, S.J., and Mulder, K. 2008. The value of coastal wetlands for hurricane protection. *Ambio* 37(4): 241–248.

Ebersole, B.A., Westerink, J.J., Bunya, S., Dietrich, J.C., and Cialone, M.A. 2010. Development of storm surge which led to flooding in St. Bernard Polder during Hurricane Katrina. *Ocean Engineering* 37: 91–103.

Feldman, A.F. 2015. Green Infrastructure Implementation in Urban Parks for Stormwater Management. (M.S. thesis, Drexel University, 2015).

Frumhoff, P.E.A., McCarthy, J.J., Melillo, J.M., Moser, S.C., and Wuebbles, D. 2007. Confronting Climate Change in the Northeast: Science, Impacts, Solutions. In: *Northeast Climate Impacts Assessment*, Union of Concerned Scientists (ed). Washington, D.C.: Union of Concerned Scientists. Accessed at: http://www.ucsusa.org/sites/default/files/legacy/assets/documents/global_warming/pdf/confronting-climate-change-in-the-u-s-northeast.pdf.

Gaffin, S.R., Rosenzweig, C., and Kong, A.Y.Y. 2012. Adapting to climate change through urban green infrastructure. *Nature Climate Change* 2(10): 704. Accessed at: http://dx.doi.org/10.1038/nclimate1685.

The Gaia Institute. 2016. Green Corridor: Lafayette Avenue Hunts Point, Bronx, NY. Accessed at: http://www.thegaiainstitute.org/Gaia/Green%20Corridor.html.

Gardner, C. 2013. U.S. Army Corps of Engineers Works After Sandy to Repair and Restore Beaches in New York Designed for Coastal Storm Risk Reduction. Accessed at: http://www.nan.usace.army.mil/Media/NewsStories/StoryArticleView/tabid/5250/Article/488052/us-army-corps-of-engineers-works-after-sandy-to-repair-and-restore-beaches-in-n.aspx.

Gedan, K.B., Kirwa, M.L., and Wolanski, E. 2011. The present and future role of coastal wetland vegetation in protecting shorelines: Answering recent challenges to the paradigm. *Climatic Change* 106: 7–29. doi:10.1007/s10584-010-0003-7.

Gold, A.R. 1991. Exxon Said to Offer Millions To Erase 1990 Harbor Spill. *The New York Times,* March 15. http://www.nytimes.com/1991/03/15/nyregion/exxon-said-to-offer-millions-to-erase-1990-harbor-spill.html.

Hayhoe, K., Wake, C.P., Huntington, T.G., Luo, L., Schwartz, M.D., et al. 2006. Past and future changes in climate and hydrological indicators in the U.S. Northeast. *Climate Dynamics* 28: 381–407.

Hunter, M. 2011. Using ecological theory to guide urban planting design: An adaptation strategy for climate change. *Landscape Journal* 30(2): 174–193.

Kosareo, L., and Ries, R. 2007. Comparative environmental life cycle assessment of green roofs. *Building and Environment* 42(7): 2606–2613.

Loder, N.M., Irish, J.L., Cialone, M.A., and Wamsley, T.V. 2009. Sensitivity of hurricane surge to morphological parameters of coastal wetlands. *Estuarine, Coastal, and Shelf Science* 84: 625–636.

Martin, D. 1994. Helping Nature Restore Life To a Shoreline Left for Dead. *The New York Times,* September 30. http://www.nytimes.com/1994/09/30/nyregion/helping-nature-restore-life-to-a-shoreline-left-for-dead.html?pagewanted=all.

Mignoni, E. 2014. Rising seas: Will the Outer Banks survive? *National Geographic.* Accessed at: http://news.nationalgeographic.com/news/special-features/2014/07/140725-outer-banks-north-carolina-sea-level-rise-climate/.

Miller, S. 2014. Evapotranspiration Potential of Green Infrastructure Vegetation. (M.S. thesis., Drexel University, 2014).

Mitsch, W.J., and Gosselink, J.G. 2000. The value of wetlands: Importance of scale and landscape setting. *Ecological Economics* 35(1): 25–33. doi:10.1016/s0921-8009(00)00165-8.

Montalto, F.A., Bartrand, T.A., Waldman, A.M., Travaline, K., Loomis, C., et al. 2012. Decentralized green infrastructure: The importance of stakeholder behavior in determining spatial and temporal outcomes. *Journal of Structure and Infrastructure Engineering* 9(12): 1–19.

Montalto, F.A., and Steenhuis, T.S. 2004. The link between hydrology and restoration of tidal marshes in the New York/New Jersey Estuary. *The Society of Wetland Scientists* 24(2): 414–425.

Natural Resources Defence Council (NRDC). 2006. Rooftops to Rivers: Green Strategies for Controlling Stormwater and Combined Sewer Overflows. Accessed at: https://www.nrdc.org/water/pollution/rooftops/rooftops.pdf.

Navarro, M., and Nuwer, R. 2012. Resisted for blocking the view. Dunes prove they blunt storms. *The New York Times* P. A1(L).

NCA (National Climate Assessment). 2013. A Review of the Draft 2013 National Climate Assessment. Accessed at: http://www.nap.edu/catalog/18322/a-review-of-the-draft-2013-national-climate-assessment.

NCA. 2014. *Climate Change Impacts in the United States. National Climate Change.* Accessed at: http://www.globalchange.gov/browse/reports/climate-change-impacts-united-states-third-national-climate-assessment-0.

NPCC (New York City Panel on Climate Change). 2013. Climate Risk Information: Observations Climate Change Projections and Maps. Accessed at: http://ccrun.org/ccrun_files/attached_files/NPCC%20Climate%20Risk%20Information%202013%20Report%206.11%20version_0.pdf.

NYC. 2007. PlaNYC: A Greener, Greater New York. Accessed at: http://www.nyc.gov/html/planyc/downloads/pdf/publications/full_report_2007.pdf.

NYC. 2009. NYC Wetlands: Regulatory Gaps and Other Threats, January 2009. Accessed at: http://www.nyc.gov/html/om/pdf/2009/pr050-09.pdf.

NYC. 2013. A Stronger, More Resilient New York. NYC Mayors Office, New York. Accessed at: http://www.nyc.gov/html/sirr/html/home/home.shtml.

NYCDCP (NYC Department of City Planning). 2011. Vision 2020: New York City Comprehensive Waterfront Plan. Accessed at: http://www.nyc.gov/html/dcp/pdf/cwp/vision2020_nyc_cwp.pdf.

NYCDCP. 2013. Coastal Climate Resilience: Urban Waterfront Adaptive Strategies.

Accessed at: http://www1.nyc.gov/assets/planning/download/pdf/plans-studies/sustainable-communities/climate-resilience/urban_waterfront_print.pdf.

NYCDEP (NYC Department of Environmental Protection). 2007. Jamaica Bay Watershed Protective Plan Executive Summary. Accessed at: http://www.nyc.gov/html/dep/pdf/jamaica_bay/vol-1-exec-summ.pdf.

NYCDEP. 2012. Department of Environmental Protection Provides Update on Repair Work at City Wastewater Treatment Plants. Accessed at: http://www.nyc.gov/html/dep/html/press_releases/12-86pr.shtml.

NYCDEP. 2014. Jamaica Bay Watershed Protection Plan. Accessed at: http://www.nyc.gov/html/dep/pdf/jamaica_bay/jbwpp_update_10012014.pdf.

NYCDEP. 2016. About watershed protection. Accessed at: http://www.nyc.gov/html/dep/html/watershed_protection/about.shtml.

Office of the Mayor. 2014. Mayor de Blasio Announces Key Resiliency Investments to Support Small Businesses and Jobs, Including New Business Resiliency Program and Major Upgrades Across Sandy-Impacted Neighborhoods [Press Release]. Accessed at: http://www1.nyc.gov/office-of-the-mayor/news/568-14/mayor-de-blasio-key-resiliency-investments-support-small-businesses-jobs-.

Peterson, G. 2000. Political ecology and ecological resilience: An integration of human and ecological dynamics. *Ecological Economics* 35(3): 323–336. doi:http://dx.doi.org/10.1016/S0921-8009(00)00217-2.

President's Council on Sustainable Development. 1999. Towards Sustainable America: Advancing Prosperity, Opportunity, and a Healthy Environment for the 21st Century. Accessed at: http://clinton2.nara.gov/PCSD/Publications/tsa.pdf/.

Rahman, M., Armson, D., and Ennos, A. 2014. A comparison of the growth and cooling effectiveness of five commonly planted urban tree species. *Urban Ecosystems* 18(2):371–389.

Rebuild by Design (RBD). 2016. Winners and Finalists. Accessed at: http://www.rebuildbydesign.org/winners-and-finalists/.

Resio, D.T., and Westerink, J.J. 2008. Modeling the physics of storm surges. *Physics Today* 61(9): 33–38. doi:10.1063/1.2982120.

Saiz, S., Kennedy, C., Bass, B., and Pressnail, K. 2006. Comparative life cycle assessment of standard and green roofs. *Environmental Science and Technology* 40(13): 4312–4316. doi:10.1021/es0517522.

Sallenger, A.H., Doran, K.S., and Howd, P.A. 2012. Hotspot of accelerated sea-level rise on the Atlantic coast of North America. *Nature Climate Change* 2(12): 884–888. doi:10.1038/nclimate1597. Accessed at: http://www.nature.com/nclimate/journal/v2/n12/abs/nclimate1597.html#supplementary-information.

Schuster, E., and Doerr, P. 2015. A Guide for Incorporating Ecosystem Service Valuation into Coastal Restoration Projects. The Nature Conservancy. Accessed at: http://www.nature.org/media/oceansandcoasts/ecosystem-service-valuation-coastal-restoration.pdf.

Sewell, B. 2011. Proposed JFK Expansion Would Harm Jamaica Bay. National Resources Defense Council Staff Blog. Accessed at: http://switchboard.nrdc.org/blogs/bsewell/proposed_jfk_expansion_would_h.html

Siegman, T. 2015. Overview of the Jamaica Bay Watershed—Helping NYC's "Jewel in the Crown" Shine. *Clean Waters*. Accessed at: https://nywea.org/clearwaters/uploads/Overview5.pdf.

Spalding, M.D., Ruffo, S., Lacambra, C., Meliane, I., Hale, L.Z., et al. 2014. The role of eco-systems in coastal protection: Adapting to climate change and coastal hazards. *Ocean and Coastal Management* 90: 50–57.

Tennant, E. 2014. 1,600 Protected Birds Killed by JFK Airport Contractors In Five Years, Records Show. *Huffington Post*. Accessed at: http://www.huffingtonpost .com/2014/05/01/jfk-birds-shot-protected-airport-port-authority_n_5246071.html.

TNC (The Nature Conservancy). 2015. Urban Coastal Resilience: Valuing Nature's Role. Case Study: Howard Beach, Queens, New York. Accessed at: https://tnc.app.box .com/s/9awez618538tf24rnu5mv73fkfa668mr/1/4092107687/34026668411/1.

Trenberth, K.E., Dai, A.G., Rasmussen, R.M., and Parsons, D.M. 2003. The changing char-acter of precipitation. *Bulletin of American Meteorological Society* 84: 1205–1217.

Tschakert, P., van Oort, B., St. Clair, A.L., and LaMadrid, A.. 2013. Inequality and trans-formation analyses: A complementary lens for addressing vulnerability to climate change. *Climate and Development* 5(4): 340–350. http://dx.doi.org/10.1080/17565529 .2013.828583.

U.S. EPA (U.S. Environmental Protection Agency). 2013. Reducing Urban Heat Islands: Compendium of Strategies. Chapter 2: Trees and Vegetation. Accessed at: https://www .epa.gov/sites/production/files/2014-06/documents/treesandvegcompendium.pdf.

U.S. EPA. 2014. United States: EPA Provides a Quarter Million Dollars to Protect Wetlands in New York. *MENA Report*. Accessed at: http://search.proquest.com/docview/163656 5389?accountid=10559.

Wamsley, T.V., Cialone, M.A., Smith, J.M., Atkinson, J.H., and Rosati, J.D. 2010. The potential of wetlands in reducing storm surge. *Ocean Engineering* 37: 59–68.

Warren, R.S. 1997. Evolution and development of tidal marshes. In: Connecticut Col-lege Arboretum Bulletin No. 34: Tidal marshes of Long Island Sound: Ecology, his-tory, and restoration. Dreyer, G.D., and Niering, W.A. (eds). Accessed at: http://camel2 .conncoll.edu/ccrec/greennet/arbo/publications/34/MAIN.htm.

Weber, T., and Wolf, J. 2000. Maryland's green infrastructure—using landscape assessment tools to identify a regional conservation strategy. *Environmental Monitoring and Assess-ment* 63(1): 265–277.

Williamson, K.S. 2003. Growing with green infrastructure. Doylestown, PA: Heri-tage Conservancy. *Environmental Monitoring and Assessment* 63(1): 265–277. doi:10.1023/A:1006416523955.

Yin, J., Overpeck, J.T., Griffies, S.M., Hu, A., Russell, J.L., and Stouffer, R.J. 2011. Different magnitudes of projected subsurface ocean warming around Greenland and Antarctica. *Nature Geoscience* 38(4): 524–528.

10

Application of Decision Science to Resilience Management in Jamaica Bay

Mitchell J. Eaton, Angela K. Fuller, Fred A. Johnson,
Matthew P. Hare, and Richard C. Stedman

This book highlights the growing interest in management interventions designed to enhance the resilience of social-ecological systems (SESs) such as the Jamaica Bay watershed. Effective management requires decision makers to anticipate how the managed system will respond to interventions (i.e., via predictions or projections), whether the focus is on managing biological processes or human behavior or (most likely) both. In systems characterized by many interacting components and high uncertainty, however, making even probabilistic predictions is difficult.

In addition to careful thinking about system dynamics, making decisions to enhance resilience in complex systems requires detailed consideration about how management objectives are specified and the selection of an analytical method used to identify the preferred action(s). Developing a clear statement of the problem(s) and articulating management objectives is an important first step and often best achieved by including input from managers, scientists, and other stakeholders affected by the decision through a process of joint problem framing (Marcot et al., 2012; Keeney et al., 1990). Decision science then provides a deliberate and transparent framework to explicitly address uncertainty and risk when making complex decisions to meet management objectives. Such a framework is particularly critical for decision makers striving to maintain resilience of desirable states when disturbances such as global change present us with such deep uncertainties (Lempert, 2002) about predicting future states.

Our goals for this chapter are to introduce the basic concepts of *decision science* and relate these in terms of possible application to *resilience management* in Jamaica Bay. Although the fields of decision science and resilience theory are complementary in many ways, integrating the two schools of thought to manage a system for resilience (see chapter 2) is not trivial. Here, we briefly highlight some of the differences in the two perspectives, but also point out potential complementarities with the belief that a productive integration of the two is achievable through a combination of careful problem framing and the application of adaptive management to address uncertainty and the dynamics of evolving systems (see below; Polasky et al., 2011a). Despite the limitations of decision science for "solving" complex resource management problems (Johnson et al., 2013; Polasky et al., 2011a), the challenges of managing for resilience require a framework that can account for diverse values, link ecological and social dynamics across scales, permit governance structures to adapt, and recognize that collective decisions made by people and institutions may affect system resilience (Folke et al., 2010). We view the application of decision science as a means to overcome such challenges by offering a logical, inclusive, and transparent means to pursue social and ecological objectives.

Decision Science in the Service of Resilience

The utility of decision science for sustaining SESs has sometimes been questioned by proponents of resilience thinking (Johnson et al., 2013; Walker and Salt, 2006), because it is held that: (1) decision problems must be framed too narrowly to make them analytically tractable; (2) the focus on a single decision maker (or even a narrow set of decision makers) optimizing a finite and unambiguously weighted set of values is unrealistic; (3) SESs are characterized by nonlinear behaviors, positive feedback loops, and surprises; (4) cross-scale interactions produce emergent behaviors that are impossible to predict; and (5) the aggressive pursuit of efficiency (e.g., via optimization) reduces system heterogeneity and therefore its adaptive capacity, thus increasing the risk of critical transitions to undesirable states.

It is indisputable that SESs are inherently complicated, but what these critiques fail to account for is that these systems, especially in urban areas, are also highly contested, with different goals held by different members of the management community and stakeholders (see chapter 6). Moreover, there remains an ongoing imperative to manage them in the context of this complexity and under high degrees of uncertainty. We would argue that there must be a principle by which to guide decision making, which at minimum includes some criterion for measuring the relative value of alternative choices, an ability to differentiate among the preferences of multiple decision makers and stakeholders, and a mechanism for selecting among alternatives.

Accepting the belief that the two schools of thought can both contribute positively to the management of complex SESs (Fischer et al., 2009; Johnson et al., 2013; Polaski et al., 2011a; Possingham and Biggs, 2012), the challenge then is to determine how to apply resilience thinking using a decision-analytic approach to successfully sustain these systems. Here we take some tentative steps in describing how decision analysis in a complex SES such as Jamaica Bay might look through the lens of "resilience thinking." We begin by defining decision science, outlining its basic components, and describing two forms of applied decision processes (structured decision making and adaptive management). Second, we suggest how considerations of managing the resilience of Jamaica Bay might be incorporated into a decision-science approach. We end by discussing opportunities and challenges in building capacity for decision science in urban watersheds.

What Is Decision Science?

Decision science emphasizes fundamental values and the deliberate trade-offs among multiple objectives that are inherent in natural resource management (Arvai et al., 2001; Keeney, 1992). Decision science explicitly structures decisions by using values (preferences) to generate objectives and evaluates the outcomes of different management actions to identify the alternative with the greatest likelihood of achieving stated objectives. Thus, a hallmark of decision science is that it is a values-based process (Keeney, 1992) in which the values and preferences of the decision makers and stakeholders guide the decision.

Values offer insights into the priorities of decision makers, highlighting conflicting interests, preferences for desired future conditions, risk tolerances, and the relative importance of different concerns. The emphasis on values helps decision makers understand whether disagreements are over uncertainties in predicted outcomes or how those outcomes are valued (Lee, 1993), and it clarifies the role for analysts and scientists in resource decision making as "honest brokers" of information rather than as policy advocates (Pielke, 2007).

Alternative processes that de-emphasize problem framing and values-based objective setting include what Gregory et al. (2012) classified as "science-based," "consensus-based," and "economic-driven" decision making. These approaches often fail to recognize that (1) social considerations must be carefully explored and quantified using social-science theories and methods so that the trade-offs among objectives are identified clearly to stakeholders (i.e., ecological science alone is unable to address what matters to society); (2) consensus approaches often fail to confront difficult trade-offs among competing objectives and, as a result, can fail to address uncertainty regarding the expected consequences of actions; and (3) economic-driven approaches tend to focus on the methods and tools used to monetize resources and ecosystem services (e.g., cost-benefit analyses) and pay less attention to uncertainty, creative alternatives, and developing an increased understanding of the problem.

An advantage of decision science is that it is a logical, integrative process that strives to structure a problem to bring clarity and insights for mutual understanding by all stakeholders, at least in part by clarifying the role that values play in decision making. The approach is becoming more commonplace in natural resource management, and is seen as contributing to robust and defensible decisions by recognizing and addressing conflicts in stakeholders' fundamental values (Allen et al., 2011; Conroy and Peterson, 2013; Gregory et al., 2012; Keeney, 1982, 1992; Walters, 1986). In the context of managing for resilience, decision science allows for integrating the ecological and social sciences to identify core values, characterizes factual information about the possible outcomes of any decision, and promotes recognition of key areas of uncertainty that reduce our confidence regarding decision outcomes.

Although specific decision-analytic approaches can vary considerably, at the most basic level all generally involve the following components (figure 10-1):

(1) *Formulating and specifying the context of the decision problem*—Framing the problem includes specifying the decision context (i.e., the set of alternatives that are appropriate to consider), identifying the decision maker(s) and stakeholders, and establishing the scope and scale of the problem (i.e., ensuring that the temporal, spatial, and organizational scales of the decision context are compatible with those of the fundamental objectives) (Keeney, 1992). Problem framing also involves efforts to identify key constraints and unacceptable outcomes, to characterize the frequency and timing of decisions to be made, and to recognize, where possible, those uncertainties that may affect the decision (Hammond et al., 1999).

(2) *Articulating the values of stakeholders and selecting criteria for quantifying progress toward achieving objectives*—Objectives are statements that describe the values or preferences of those involved in, or affected by, the decision-making process. Eliciting objectives from stakeholders and decision makers allows for the incorporation of multiple views, highlights potential conflicts among objectives, engenders greater acceptance of the eventual decision, and begins the process of building common understanding among participants. Appropriate performance criteria or attributes must be selected for each objective. Attributes clarify and operationalize the meaning of objectives (i.e., reduce misinterpretation), provide a means for evaluating the consequences of alternatives with respect to the objectives, and allow quantification of management success (i.e., via monitoring).

(3) *Identifying feasible management actions*—Alternative management actions (choices) should be created from a shared vision of objectives among key stakeholders and decision makers and how they might be achieved (Keeney, 1992). Approaches that identify alternatives prior to an explicit statement of objectives can lead to suboptimal decisions because the alternatives were not crafted for the purpose of trying to achieve agreed-upon objectives. Thus, one is limited to asking the question, "is this alternative acceptable?," rather than "how well does this meet our objectives?" It is important that the set of

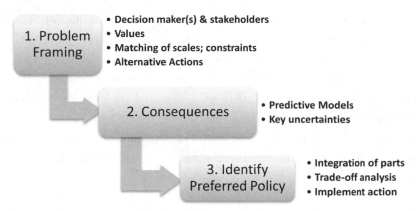

Figure 10-1. Primary components of a decision-analytic approach to transparent, defensible decision making. Courtesy of Mitch Eaton of the U.S. Geological Survey.

alternatives (1) be sufficient to describe all the ways that the objectives could be achieved (Keeney, 1992), (2) can be properly evaluated, (3) includes actions that differ sufficiently to present decision makers with a real choice, and (4) allows for learning to occur when decisions are iterated.

(4) *Evaluating potential outcomes or consequences of implementing the actions relative to stated objectives*—Predicting the consequences of all proposed management alternatives (including a decision to do nothing) in terms of the objectives is required to understand and compare potential actions to evaluate trade-offs and make a decision. Thus, either quantitative or qualitative models of system dynamics (as reviewed in chapter 8) are needed as means to predict the response to management interventions.

(5) *Evaluating the trade-offs associated with taking one management action over another*—Trade-offs are inevitable in decision making. Optimization methods are used to identify the best decision choice to achieve the objectives, assuming that the relative importance of objectives can be agreed upon. Even for single-objective problems, the risk attitude of a decision maker represents a competing value between the potential benefits of choosing a course of action and the negative impacts if that action results in an undesirable outcome. For example, a risk-averse decision maker may trade an alternative with the highest expected net benefit for one with a lower expected benefit if the latter is more certain (i.e., less likely to produce an unwanted outcome). A variety of quantitative tools are available to explicitly evaluate trade-offs and allow decision makers and stakeholders to understand how much of one objective must be traded off or reduced to achieve gains in another objective. A comprehensive description of the analytical tools used in decision science is beyond the scope of this chapter, but they include utility theory, multicriteria decision analysis (Belton and Stewart, 2002), classic optimization, probability theory, risk assessment, Monte Carlo simulation, stochastic dynamic programming (Conroy and Moore,

2001), Markov decision processes (Williams, 2009), expected value of information (Runge et al., 2011; Williams et al., 2011), adaptive management (McGowan et al., 2011; Tyre et al., 2011; Williams and Johnson 1995), and many others.

(6) *Deciding and taking action*—The steps of a decision-science approach described above should result in identifying a preferred action that then can be implemented. As understanding of the system evolves and uncertainties are reduced, the management decision may change through time. The integration of learning (specifically, the reduction of uncertainty) in the decision-making process makes this process adaptive (i.e., adaptive management).

Applying Decision Science: Structured Decision Making and Adaptive Management

Structured decision making and adaptive management are two formalized (and related) approaches for organizing and applying the principles of decision science to the process of making decisions (Gregory et al., 2012; Conroy and Peterson, 2013). Both of these processes closely follow the six components described above and offer guidance for additional considerations, including the roles of scientists and stakeholders (Robinson and Fuller, in press), an emphasis on coproduction and transparency (McNie, 2007), and the imperative for targeted monitoring in decision making (Lyons et al., 2008). Adaptive management extends the basic structured decision-making process to address decisions that contain substantial uncertainty and reoccur over space or time, thus providing an opportunity to learn about system dynamics and adapt decisions to reflect reductions in uncertainty.

Adaptive management is relevant to managing for resilience because it explicitly incorporates uncertainty associated with imperfect understanding of the system, or incomplete control of management actions, and can identify which sources of uncertainty are most likely to be relevant to the particular decision (Allen and Gunderson, 2011; Johnson and Williams, 2014; Runge et al., 2011). Uncertainty in system dynamics, or the functioning of ecosystems, is addressed by comparing alternative predictions from competing models (representing critical uncertainties) with observations of how a system responds after management is implemented. Such a comparison results in an updating of the support for each hypothesis relative to its performance (i.e., the credibility or "weight" for each model is modified using Bayes theorem; Williams, 1996). The updating of model support then provides greater confidence in composite model predictions during the next decision iteration (Johnson et al., 2015).

Well-designed monitoring programs allow for such evaluation, with a focus on gaining information to help inform management decisions (i.e., learning about system states), evaluating the effectiveness of management actions on achieving objectives, and providing a feedback loop for learning about the system (Lyons et al., 2008). Adaptive management encourages learning over time in terms of both the behavior of ecosystems and the effectiveness of management actions in achieving specified objectives. This approach represents a formalized mode of learning, via the reduction of key uncertainties that affect the decision, rather than an ad hoc process of trial and error.

Although adaptive management was primarily designed to focus learning on understanding system dynamics and responses to management, the application of decision science accommodates multiple modes and scales of learning. Learning at different levels has been described in the management theory literature as single-, double-, and triple-loop cycles (figure 10-2; Pahl-Wostl, 2009).

Single-loop learning is characterized in adaptive management by incremental improvements in management based on initial assumptions (e.g., system models) and an established decision-making framework. Double-loop learning occurs when initial assumptions and conditions are called into question and leads to a re-examination or reframing of management objectives, available management alternatives, or hypotheses of system behavior. Finally, triple-loop learning constitutes a re-examination and transformation of social or institutional constraints and processes that may be restricting the ability of governance structures to adapt as resource management needs evolve in a dynamic social context. Just as the single-loop cycle permits the use of observations of the

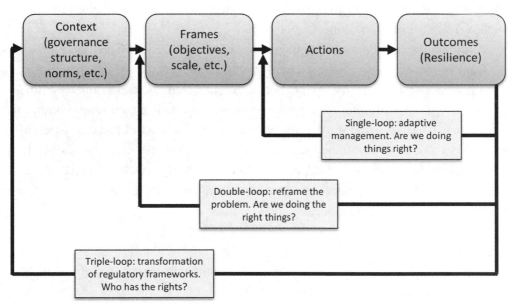

Figure 10-2. Hierarchical levels of learning in a triple-loop learning cycle. Different time- and governance scales should be accounted for when incorporating learning to promote adaptation and innovation. (Adapted from Pahl-Wostl, 2009).

system to improve our understanding of socio-ecological dynamics, observations of emergent political processes following management implementation provide opportunities to learn about and adapt governance systems (Pahl-Wostl, 2009). Indeed, the flexibility to transition governance systems has been cited as a necessary condition for using adaptive management to address climate change (Cosens et al., 2014). Although few adaptive management programs have been in place long enough to invoke double- and triple-loop learning cycles, one long-term program has demonstrated that learning is not limited to reducing structural uncertainty, and that adaptation to evolving objectives and assumptions, as well as forms of governance, can be included in a decision-analytic framework (Johnson et al., 2015).

Decision Analysis for Resilience Management in Jamaica Bay

Decision science has been widely used in business and government decision making (Keefer et al., 2004), but its application to problems in natural resource management has mostly been a phenomenon of the last two decades (see review of case studies in Conroy and Peterson, 2013). Examples range from management of recreational fisheries (Irwin et al., 2008; Peterson and Evans, 2003) and threatened and endangered species conservation (Gregory and Long, 2009; Tyre et al., 2011) to management of invasive species (Blomquist et al., 2010; Runge, 2011), water use planning (Gregory and Failing, 2002), and estuarine habitat management (Robinson and Jennings, 2012). Many of these are similar to the types of decisions that managers face in Jamaica Bay, and include applications embedded in SESs (Failing et al., 2007; Runge, 2011).

In the following sections we highlight considerations likely to be relevant for management of Jamaica Bay within a decision-science framework. These ideas are presented as broadly applicable to many types of decisions and system dynamics, and we contend that explicit incorporation of concepts from resilience thinking is compatible within this framework. We believe that the desire to maximize the resiliency of Jamaica Bay to drivers of global change is a sensible goal and consistent with a decision-analytic approach to management, but assert that resilience thinking itself can benefit from the clarification of terms, model variables, metrics, and other problem elements that are essential to any systematic decision-making process.

Problem Framing

Properly specifying or framing the problem is the essential first step in a decision-analytic approach. It may be common for a manager or stakeholder in the Jamaica Bay watershed to express a variety of values, including, for example, a desire to improve water quality, protect sensitive wildlife and habitats, or increase public access to the waterfront. Operationalizing such statements, however, into unambiguous objectives with clearly defined performance metrics, management actions that are of an appropriate scale, and a transparent means to evaluate trade-offs may not be so obvious.

A key difficulty for decision analysts is the "problem of fit" among the different components of SESs (Folke et al., 2007). Scale mismatches "occur when the scale of environmental variation and the scale of the social organization responsible for management are aligned in such a way that one or more functions of the social-ecological system are disrupted" (Cumming et al., 2006, p. 3). In other words, scale mismatches have to do with the relationship between the scales of a decision maker's ability to affect change in a system and the scale of environmental and/or social issues of concern. For example, the National Park Service (NPS) is a decision maker concerned about the integrity of salt marshes in Jamaica Bay, which are themselves strongly influenced by water quality, but water quality is largely controlled by decision makers in New York City and the increasing population there (which in turn is driven by other forces, such as migration rates, economic development, etc.). Therefore, it may be beyond the control of the NPS to meet objectives concerned with salt marsh integrity. Recognizing this, the most recent management plan for Gateway National Recreation Area (NPS, 2014), which includes Jamaica Bay, identifies coordination among agencies, including the city of New York and other research groups such as the Science and Resilience Institute at Jamaica Bay, as an important means for improving natural resource management in the context of a multipurpose urban park.

The success of restoring and sustaining Jamaica Bay depends, therefore, on many decisions made by a variety of decision makers at different scales and under different sets of structural constraints. Although overlapping governance structures add complexity, successful applications of decision science to systems managed by multiple decision makers are becoming more common (McGowan et al., 2015). A key challenge will be to understand the linkages among decisions by different actors and how they might erode or enhance resilience of desirable social-ecological states.

Resilience proponents also emphasize that environmental and social drivers interact across scales. To address such cross-scale interactions, analysts must consider both smaller-faster and larger-slower processes than the focal scale of decisions (Gunderson, 2001). These processes help define the "noise" (i.e., uncontrolled variation) in the case of the former, and the constraints in the case of the latter, that should be explicitly considered in the decision analysis (Johnson et al., 2013; Beck, 2010). An example would be that large-scale climatic and other physical forces (anthropogenic or otherwise) drive sea level rise which, in turn, imposes constraints on the actions available to decision makers in Jamaica Bay. On the other hand, water quality in Jamaica Bay is a product of fine-scale processes of land use, precipitation, and runoff. The resulting spatial and temporal variability in water quality must be summarized (e.g., averaged) at the scale of the decision-making process to help predict the consequences (including how these consequences are subjectively experienced and responded to by stakeholders) of alternative treatment options (i.e., small-scale variation is treated as "statistical noise" that must be accounted for at the scale of the decision).

Objective Setting

Identifying and agreeing on social-ecological objectives in Jamaica Bay is extraordinarily challenging given the diversity of the system, and the broad range of factors represented and actors involved (as discussed in chapter 5). The very notion of what constitutes a desirable state is potentially problematic when we consider the diversity of stakeholder preferences (Davidson, 2010). Different social actors perceive the state of the system differently, value these states differently, and have different capacities to act on these perceptions and preferences, leading to a diversity of potential behavioral responses (Stedman, 2015). This reinforces earlier assertions in this chapter about decision making explicitly engaging values and objectives, rather than attempting to portray the process as value-free.

Some objectives, such as for supporting services (e.g., ecosystem processes, diversity maintenance, and disturbance cycles), control the structure and functioning of ecosystems, yet are often overlooked because they are not directly valued by society (Chapin et al., 2009), nor is management particularly well positioned to address them. Ecosystem processes particularly relevant for Jamaica Bay include sediment deposition and transport, freshwater supply and hydrodynamic flushing rate, and carbon and nutrient (especially nitrogen) cycling (Waldman, 2008). It is these slowly changing variables that are most likely to affect the "stability landscape," and thus the degree of resilience of desirable and undesirable stability regimes (Scheffer, 2009). Analogous social-economic drivers include changes in population, wealth, inequality, environmental access, and awareness/concern. Using an understanding of these processes to develop objectives for decision making can help avoid a focus on more short-term, myopic objectives that, although well intended, can eventually erode resilience (Holling and Meffe, 1996; Walker and Salt, 2006).

Biodiversity—indeed, diversity in general—is often touted as an objective important for maintaining the capacity of systems to adapt to disturbance (Levin, 1999). Some evidence suggests, however, that it is not species richness per se that enhances system resilience, but the way in which species interact with each other and with their environment (Peterson et al., 1998). Social diversity, ranging from individuals to institutions, functions much the same way as biological diversity (Stedman et al., 2012). For example, economic diversity as expressed by a broad range of economic sectors does not automatically confer resilience if these sectors respond in similar fashion to system shocks. In these cases, the focus should be on ecological and socioeconomic processes, not pattern, as pattern can be an inadequate surrogate for process. Thus, objectives for resilience-based management might be to maximize the diversity of functional entities while maintaining some degree of redundancy within groups. Here, redundancy refers to entities with overlapping function, but that respond to environmental, economic, or sociological variation differently (Walker, 1992, 1995). An important implication is that not all system components contribute equally to resilience, and objectives (and their weights) can be specified to capture these differences in values. Species are adapted to the disturbance regimes that shaped their evolutionary history, and decisions to alter these regimes for short-term benefits (e.g., flood control) produce conditions to which species are less well adapted (as well as conditions under which exotic invaders can flourish). These structural changes can in turn erode resilience and increase the risk of undesirable regime shifts (Chapin et al., 2009). Therefore, objectives that focus on the *variability* in system properties such as hydrodynamic flushing rate, rather than their *expected* value, are more likely to be appropriate for resilience-based management (Moore and McCarthy, 2010).

Beyond a focus on processes and features that enhance resilience (or promote transformation), the temporal dimension of decision making also plays an important role in sustaining SESs. Overly discounting the future or other forms of myopic decision making that focus on short-term returns can degrade sustainability because of the legacy of decisions made in the present. Myopic decision making also severely lowers the value of reducing uncertainty, which helps to improve decision making in the future (i.e., adaptive management) (Hauser and Possingham, 2008; Moore et al., 2008). Social-ecological memory, another temporal component of learning, is likewise seen as key for enhancing resilience, and has been applied especially to urban systems (Barthel et al., 2010).

Finally, the mathematical expression of how objectives and outcomes are valued (i.e., objective functions) can be framed to account for improbable, but highly undesirable, outcomes. Non-linear utility curves can be used to represent the risk attitude of stakeholders so that the selection of management alternatives that might result in undesirable outcomes is avoided (McGowan et al., 2015). A key requirement in the context of resilience thinking is the ability to model the processes driving regime shifts (see "Consequences" section below), though in most cases the understanding of such processes will be limited, and often is only realized after a regime shift. This uncertainty has important implications for the conduct of adaptive management.

Intuition suggests that when resilience is low and the costs associated with undesirable states are high, system probing or experimentation to facilitate learning is unlikely to be prudent (Allen and Gunderson, 2011; Gunderson, 1999). Indeed, application of adaptive optimization (Williams, 1996; Williams, 2001) likely will produce management strategies that minimize the probability of moving to system states associated with high costs, unless expected learning is sufficient (and the system resilient enough) to recoup the costs over the time frame of decision making.

A productive line of inquiry thus involves understanding how various sources and degrees of uncertainty in the mechanics of regime shifts influence optimal prescriptions for adaptive management. Methods of decision analysis that focus on variability in objective returns and on robust decision making are more likely to be relevant in these cases than classic methods that focus on maximizing expected values (more on this in the "Trade-offs" section below).

Alternative Actions

Resilience thinking leads to a number of suggestions about how alternative actions should be formulated for decision analysis. To the extent possible, actions should be targeted at "slow" variables that are the principal determinants of system resilience and at those that are reversible (e.g., "soft" versus "hard" solutions to sea level rise; Cundill et al., 2012). Actions designed to stabilize the system or its returns—i.e., "lock-ins" into a particular state—should be avoided (Holling and Meffe, 1996). Suites of actions can be bundled into portfolios, representing a realistic diversity of activities that can be implemented together to help reduce vulnerability, enhance adaptive capacity, and navigate desired transformations in SESs (Adger et al., 2005; Chapin et al., 2009). This was the approach taken in the Gateway management plan (NPS, 2014), in which one portfolio of management policies was chosen over two alternatives (including a status quo option), emphasizing diversified public access and recreational opportunities over concentrated access and greater preservation effort.

Consequences

Constructing a predictive model is an essential aspect of any systematic approach to decision making. A resilience-based perspective, however, emphasizes the difficulty of making even probabilistic predictions because of the need to generalize from limited experience, a lack of understanding of mechanisms that can generate extreme events, the presence of "deep" uncertainty (Carpenter et al., 2009; Peterson et al., 2003), and the local "surprise" that accompanies the difficulty of predicting human response (Holling, 1986).

Scenario planning has been advocated as an alternative decision-making strategy (Peterson et al., 2003; Polasky et al., 2011a), even though plausible scenarios ideally arise from a (possibly implicit) process of model building (though the models need not be mechanistic or provide a stochastic structure for future scenarios). Scenarios are described as coherent, internally consistent, and plausible descriptions of possible future states and are often constructed to reflect, as best as possible, the range of uncertainty in outcomes (Mahmoud et al., 2009). In general, scenarios represent credible hypotheses about how the future might unfold, often with a description of the pathways (including policy interventions) leading to these potential, but not equally likely, outcomes. For the purposes of decision analysis, we suggest that scenarios should be treated the same as any other outcome of traditional modeling. Thus, we view a "model" in its broadest sense as any sort of state and action-dependent outcome or as an algorithm for generating such a prediction, projection, or scenario.

Proponents of resilience thinking emphasize the risk of catastrophic regime shifts in SESs (e.g., the conversion of scrub lands to desert, or the shift from low-turbidity estuaries dominated by macrophytes to high-turbidity systems dominated by algae). But not all regime shifts in SESs are catastrophic, and not all systems exhibit pronounced hysteresis (hysteresis refers to the difficulty of returning a system to a more desirable state after an undesirable regime shift; Scheffer, 2009). In systems where gradual change is typical or hypothesized, classic decision analysis and its variants are appropriate and remain valuable tools for resource management and conservation. Decision analysis can also be useful for models involving multiple regimes. For example, simple models have been used to demonstrate how optimal management differs under various assumptions about the nature of regime changes (Polasky et al., 2011b). Precautionary management may be best if a potential regime shift changes system dynamics, and if management affects the probability of a regime shift. The relationship between such dynamics and management decisions has been explored theoretically using the concept of ecological and decision thresholds (Eaton et al., 2014).

The key requirement for the existence of alternative stability regimes is one or more positive feedback loops (i.e., those that are reinforcing rather than stabilizing) with sufficient strength to generate a stability landscape (Scheffer, 2009). The resilience of alternative states is controlled by external factors (usually "slow" variables) and can change over time. Known mechanisms that can generate alternative stable states include overharvesting in the presence of an Allee effect, changes in trophic structure, fragmentation of landscapes, interspecific competition (especially as it applies to invasions by exotics), and transmission of disease (Scheffer, 2009). A simple example of how positive feedback loops in Jamaica Bay might contribute to alternative stability regimes is this: there is a positive feedback between oysters and phytoplankton—low abundance of oysters (filter feeders) increases the abundance of phytoplankton, which in turn increases the amount of organic matter and decomposition, which in turn further decreases the levels of oysters (through hypoxia). Similarly, with respect to social phenomena, we often see positive feedbacks between variables such as crime rates, out-migration (to avoid crime), and the erosion of services (such as policing) through declining tax revenues. Such processes can lead to alternative stability regimes that—similar to the ecological example above—can be difficult to escape (i.e., hysteresis). This theoretical basis for understanding the dynamics of complex systems can serve as a productive foundation for developing testable hypotheses (models) or scenarios to guide decision making and system monitoring under an adaptive management framework.

Trade-offs (Optimization)

Jamaica Bay represents a system characterized by decision problems that are both dynamic and burdened by significant uncertainty. In such cases, time- and state-dependent management policies can be solved using dynamic optimization methods such as stochastic dynamic programming (Marescot et al., 2013). A key advantage of dynamic optimization methods is their ability to specify optimal decisions for *possible* future system states rather than *expected* future states. In practice this makes dynamic optimization appropriate for systems that behave stochastically, absent assumptions about the system remaining in a desired equilibrium or about the production of a constant stream of resource returns.

From a resilience perspective, optimization differs from classical decision approaches in some important ways. Although it is often optimal from an efficiency perspective to hold a system near a bifurcation (tipping) point, this also reduces the system's resilience and leaves it vulnerable to shocks or other sources of uncertainty. Rather than focusing on efficiency, a cautious approach might focus on minimizing the likelihood of regime shift or achieving at least a minimum level of performance across the greatest range of uncertainty. Characterizing objectives and management benefits in this manner may be more appropriate, particularly if there is uncertainty about the distance from the tipping point and if the costs of collapse are high (Scheffer, 2009).

Classical decision-science approaches may also be less appropriate when there is deep uncertainty about the state of the system and its dynamics. In these cases, it is no longer is meaningful to optimize based on an average of performance values, because there is no known distribution on which to base the averaging. A different criterion is needed to guide decision making. One such candidate is robust decision making. Here the idea is not to maximize a measure of management performance (utility), but rather to produce values exceeding some specified lower limit over as large a range of possible system behaviors as possible (Ben-Haim, 2001; Regan et al., 2005). Robust decision making involves the choice of an action that will maximize the range of system behaviors for which the expected return for every system behavior in that range will be "good enough." This shifts the focus from maximizing expected return to maximizing coverage of an acceptable value. The operative question is, "how wrong can one be about the system behaviors and still produce an adequate return?" (commonly referred to as "info-gap"; Ben-Haim, 2001).

Another trade-off approach potentially useful in resilience-based management is the notion of Pareto optimality (Kennedy et al., 2008). Pareto-optimal solutions are those in which one set of values cannot be improved upon without a reduction in other values (Bishop, 1993; Polasky et al., 2008). Although there can be no single, optimal solution (because there is no agreement on objectives or how they are weighted), Pareto-optimal solutions provide a basis for negotiating a solution among stakeholders by first ruling out solutions that do not perform well on any of the objectives.

Building Capacity for Decision-Analytic Resilience Management in Jamaica Bay

In this chapter, we have described the potential application of decision science to resilience management in Jamaica Bay and elsewhere. It is our belief that the concepts of resilience thinking can be integrated into an adaptive management framework and, conversely, that decision science offers a practical means to bring clarity and provide sensible guidance to managing for resilience and adaptive capacity.

How is resilience-based adaptive management possible in an urban watershed setting with many overlapping jurisdictions, high levels of uncertainty, and potentially conflicting value systems? Management in an urban watershed may be particularly challenging due to high and competing demands for natural resources and the dissociation of decision makers from environmental feedback (Folke, 2006). The first requirement is for an integrated approach to system management, but in this domain the roles of scientists, decision makers, and stakeholders have traditionally been largely independent or, at the very least, poorly integrated. Science has often been viewed as an autonomous process, to the extent that institutional separation of science from policy has been commonplace (Mills and Clark, 2001; McNie, 2007). Scientists conduct investigations to reduce uncertainty and better understand system dynamic processes, often with a desire that their results will be useful for management purposes. However, because these studies are typically conceived and implemented without the input of managers—and often with different types of science working in isolation from each other—they are commonly focused at a scale or on variables that are unsuitable or uninformative for decision making. Many scientists also believe that management of system elements, whether ecological or social, should be science-based rather than values-focused, which leads to conflicts between scientists and stakeholders. Conversely, managers and policy makers often do not communicate with scientists at an early stage in framing decisions, which limits the ability of researchers to understand the managed resource. Finally, neither managers nor scientists routinely seek to include the public or other stakeholders when formulating management objectives and alternatives.

We believe that the most effective integration of management and science begins at the very outset of defining the management process, and that codevelopment of all or most of the components of the decision problem by decision makers, scientists, and stakeholders increases accountability in decision making. This in turn engenders greater ownership, creates perceptions of transparency and fairness, and can reduce conflict in situations of multiple, competing objectives (Lauber and Knuth, 1997). This approach may also foster conditions leading to greater socio-ecological memory, institutional learning, social networks, and other traits of a resilient and adaptive governance system (Folke, 2006).

An integrated approach to SES management aims to be collaborative, focused on objectives, and based on scientific information and evidence. Scientific research and management is collaborative when both scientists and the stakeholders responsible for management are involved in the decision process. This means that dialogue is necessary at each step of the process. Scientists should seek input from stakeholders when determining what research agendas in a particular system will best contribute to management. Likewise, stakeholders should provide feedback to scientists on what information and data are needed to best inform the decision-making process. Together, stakeholders and researchers can set an agenda for research that is focused on specific objectives designed to yield the most useful information for management decisions. Greater integration is needed throughout the process, including sharing data and results, communicating about the decision process, developing a monitoring policy, and supplementing system understanding with new scientific research and new policies when appropriate.

Recent developments in New York and Jamaica Bay suggest a growing capacity for integrative management. For example, legislation was passed in 2006 to implement ecosystem-based management of all New York State marine resources, demonstrating at the state level a commitment to moving from single-species management to a systems-based approach that integrates human-value systems (New York Ocean and Great Lakes Ecosystem Conservation Act, 2006). In Jamaica Bay, even before Hurricane Sandy in 2012, the National Park Service and City of New York recognized the need for a bridging organization that could channel scientific expertise into a more integrated and productive management process (NPS, 2014). It was through their request for statements of interest, and with the benefit of established bridging organizations such as the Rockefeller Foundation and Jamaica Bay Conservancy, that the Science and Resilience Institute at Jamaica Bay was conceived and formed with the mission described in chapter 1. Under the auspices of this institute, there are three active councils: scientific, public agency, and citizen stakeholder groups (chapter 12). Recent participation in the public agency council by decision makers from over a dozen city, state, and federal authorities is encouraging because their overlapping jurisdictions and sometimes conflicting missions need to be accommodated for a systems approach to management. Collectively, these recent developments elevate the potential for an integrated approach to managing resilience in Jamaica Bay, which can be enhanced by the use of decision science.

References

Adger, W.N., Hughes, T.P., Folke, C., Carpenter, S.R., and Rockström, J. 2005. Social-ecological resilience to coastal disasters. *Science* 309(5737): 1036–1039.

Allen, C.R., Fontaine, J.J., Pope, K.L., and Garmestani, A.S. 2011. Adaptive management for a turbulent future. *Journal of Environmental Management* 92: 1339–1345.

Allen, C.R., and Gunderson, L.H. 2011. Pathology and failure in the design and implementation of adaptive management. *Journal of Environmental Management* 92: 1379–1384.

Arvai, J.L., Gregory, R., and McDaniels, T.L. 2001. Testing a structured decision approach: Value-focused thinking for deliberative risk communication. *Risk Analysis* 21(6): 1065–1076.

Barthel, S., Folke, C., and Colding, J. 2010. Social–ecological memory in urban gardens— retaining the capacity for management of ecosystem services. *Global Environmental Change* 20(2): 255–265.

Beck, U. 2010. Climate for change, or how to create a green modernity? *Theory, Culture and Society* 27(2–3): 254–266.

Belton V., and Stewart, T.J. 2002. *Multiple Criteria Decision Analysis: An Integrated Approach.* Norwell, MA: Kluwer Academic Publishers.

Ben-Haim, Y. 2001. *Information Gap Decision Theory: Decisions Under Severe Uncertainty.* San Diego, CA: Academic Press.

Bishop, R. 1993. Economic efficiency, sustainability, and biodiversity. *Ambio* 22(2): 69–73.

Blomquist, S.M., Johnson, T.D., Smith, D.R., Call, G.P., Miller, B.N., et al. 2010. Structured decision-making and rapid prototyping to plan a management response to an invasive species. *Journal of Fish and Wildlife Management* 1: 19–32.

Carpenter, S.R., Folke, C., Scheffer, M., and Westley, F. 2009. Resilience: Accounting for the noncomputable. *Ecology and Society.* 14(1):13.

Chapin, F.S., Kofinias, G.P., and Folke, C. (eds). 2009. *Principles of Ecosystem Stewardship: Resilience-Based Natural Resource Management in a Changing World.* New York, NY: Springer.

Conroy, M.J., and Moore, C.T. 2001. Simulation models and optimal decision making in natural resource management. In: *Modeling in Natural Resource Management: Development, Interpretation, and Application.* Shenk, T., and Franklin, A. (eds). Washington, D.C.: Island Press.

Conroy, M.J., and Peterson, J.T. 2013. *Decision Making in Natural Resource Management: A Structured, Adaptive Approach.* West Sussex, UK: John Wiley & Sons, Ltd.

Cosens, B., Gunderson, L., Allen, C., and Benson, M. 2014. Identifying legal, ecological and governance obstacles, and opportunities for adapting to climate change. *Sustainability* 6: 2338–2356.

Cumming, G.S., Cumming, D.H.M., and Redman, C.L. 2006. Scale mismatches in social-ecological systems: Causes, consequences, and solutions. *Ecology and Society* 11(1): 14.

Cundill, G., Cumming, G.S., Biggs, D., and Fabricius, C. 2012. Soft systems thinking and social learning for adaptive management. *Conservation Biology* 26: 13–20.

Davidson, D.J. 2010. The applicability of the concept of resilience to social systems: Some sources of optimism and nagging doubts. *Society and Natural Resources* 23(12): 1135–1149.

Eaton, M.J., Martin, J., Nichols, J.D., McIntyre, C., McCluskie, M.C., et al. 2014. Application of threshold concepts in natural resource decision making. In: *Application of Threshold Concepts in Natural Resource Decision Making.* Guntenspergen, G.R. (ed). 67–86. New York, NY: Springer.

Failing, L., Gregory, R., and Harstone, M. 2007. Integrating science and local knowledge in environmental risk management: A decision-focused approach. *Ecological Economics* 64: 47–60.

Fischer, J., Peterson, G.D., Gardner, T.A., Gordon, L.J., and Fazey, I., et al. 2009. Integrating resilience thinking and optimisation for conservation. *Trends in Ecology & Evolution* 24: 549–554.

Folke, C. 2006. Resilience: The emergence of a perspective for social–ecological systems analyses. *Global Environmental Change* 16: 253–267.

Folke, C., Carpenter, S., Walker, B., Scheffer, M., Chapin, F.S., and Rockström, J. 2010. Resilience thinking: Integrating resilience, adaptability and transformability. *Ecology and Society* 15(4): 20.

Folke, C., Pritchert, J., Berkes, F., Colding, J., and Svedin, U. 2007. The problem of fit between ecosystems and institutions: Ten years later. *Ecology and Society* 12(1): 30.

Gregory, R., and Failing, L. 2002. Using decision analysis to encourage sound deliberation: Water use planning in British Columbia, Canada. *Journal of Policy Analysis and Management* 21: 492–499.

Gregory, R., Failing, L., Harstone, M., Long, G., McDaniels, T., and Ohlson, D. 2012. *Structured Decision Making: A Practical Guide to Environmental Management Choices*. West Sussex, UK: John Wiley & Sons, Ltd.

Gregory, R.S., and Long, G. 2009. Using structured decision making to help implement a precautionary approach to endangered species management. *Risk Analysis* 29: 518–532.

Gunderson, L.H. 1999. Resilience, flexibility and adaptive management—antidotes for spurious certitude? *Conservation Ecology* 3(1): 7.

Gunderson, L.H. 2001. *Panarchy: Understanding Transformations in Human and Natural Systems*. Washington, D.C.: Island Press.

Hammond, J.S., Keeney, R.L., and Raiffa, H. 1999. *Smart Choices: A Practical Guide to Making Better Life Decisions*. New York: Broadway Books.

Hauser, C.E., and Possingham, H.P. 2008. Experimental or precautionary? Adaptive management over a range of time horizons. *Journal of Applied Ecology* 45: 72–81.

Holling, C.S. 1986. Resilience of ecosystems: Local surprise and global change. In: *Sustainable Development of the Biosphere*. Clark, W.C., and Munn, R.E. (eds). 292–317. Cambridge (UK): Cambridge University Press.

Holling, C.S., and Meffe, G.K. 1996. Command and control and the pathology of natural resource management. *Conservation Biology* 10(2): 328–337.

Irwin, B.J., Wilberg, M.J., Bence, J.R., and Jones, M.L. 2008. Evaluating alternative harvest policies for yellow perch in southern Lake Michigan. *Fisheries Research* 94: 267–281.

Johnson, F.A., Boomer, G.S., Williams, B.K., Nichols, J.D., and Case, D.J. 2015. Multilevel learning in the adaptive management of waterfowl harvests: 20 years and counting. *Wildlife Society Bulletin* 39: 9–19.

Johnson, F.A., and Williams, B.K. 2014. A decision-analytic approach to adaptive resource management. In: *Adaptive Management of Natural Resources in Theory and Practice*. Allen, C.R., and Garmestani, A.S. (eds). 61–84. New York, NY: Springer Publishing Co.

Johnson, F.A., Williams, B.K., and Nichols, J.D. 2013. Resilience thinking and a decision-analytic approach to conservation: Strange bedfellows or essential partners? *Ecology and Society* 18(2).

Keefer, D.L., Kirkwood, C.W., and Corner, J.L. 2004. Perspective on decision analysis applications. *Decision Analysis* 1(1): 4–22.

Keeney, R.L. 1982. Decision analysis: An overview. *Operations Research* 30(5): 803–838.

Keeney, R.L. 1992. *Value-Focused Thinking: A Path to Creative Decisionmaking*. Cambridge, MA: Harvard University Press.

Keeney, R.L., von Winterfeldt, D., and Eppel, T. 1990. Eliciting public values for complex policy decisions. *Management Science* 36: 1011–1030.

Kennedy, M.C., Ford, E.D., Singleton, P., Finney, M., and Agee, J.K. 2008. Informed multi-objective decision-making in environmental management using Pareto optimality. *Journal of Applied Ecology* 45: 181–192.

Lauber, T., and Knuth, B. 1997. Fairness in moose management decision-making: The citizens' perspective. *Wildlife Society Bulletin* 25: 776–787.

Lee, K.N. 1993. *Compass and Gyroscope: Integrating Science and Politics for the Environment*. Washington, D.C.: Island Press.

Lempert, R.J. 2002. A new decision sciences for complex systems. *PNAS* 99: 7309–7313.

Levin, S. 1999. *Fragile Dominion: Complexity and the Commons*. Reading, MA: Perseus Books.

Lyons, J.E., Runge, M.C., Laskowski, H.P., and Kendall, W.L. 2008. Monitoring in the context of structured decision-making and adaptive management. *Journal of Wildlife Management* 72: 1683–1692.

Mahmoud, M., Liu, Y., Hartmann, H., Stewart, S., Wagener, T., et al. 2009. A formal framework for scenario development in support of environmental decision-making. *Environmental Modelling and Software* 24: 798–808.

Marcot, B.G., Thompson, M.P., Runge, M.C., Thompson, F.R., McNulty, S., et al. 2012. Recent advances in applying decision science to managing national forests. *Forest Ecology and Management* 285: 123–132.

Marescot, L., Chapron, G., Chadès, I., Fackler, P.L., Duchamp, C., et al. 2013. Complex decisions made simple: A primer on stochastic dynamic programming. *Methods in Ecology and Evolution* 4: 872–884.

McGowan, C.P., Lyons, J.E., and Smith, D.R. 2015. Developing objectives with multiple stakeholders: Adaptive management of horseshoe crabs and red knots in the Delaware Bay. *Environmental Management* 55: 972–982.

McGowan, C.P., Smith, D.R., Sweka, J.A., Martin, J., Nichols, J.D., et al. 2011. Multispecies modeling for adaptive management of horseshoe crabs and red knots in the Delaware Bay. *Natural Resource Modeling* 24: 117–156.

McNie, E.C. 2007. Reconciling the supply of scientific information with user demands: An analysis of the problem and review of the literature. *Environmental Science and Policy* 10: 17–38.

Mills, T.J., and Clark, R.N. 2001. Role of research scientists in natural resource decision-making. *Forest Ecology and Management* 153: 190–198.

Moore, A.L., Hauser, C.E., and McCarthy, M.A. 2008. How we value the future affects our desire to learn. *Ecological Applications* 18(4): 1061–1069.

Moore, A.L., and McCarthy, M.A. 2010. On valuing information in adaptive-management models. *Conservation Biology* 24(4): 984–993.

New York Ocean and Great Lakes Ecosystem Conservation Act. 2006. S-8380, A-10584B. Accessed at: http://www.dos.ny.gov/opd/programs/pdfs/ECL_Article14.pdf.

NPS (National Park Service). 2014. Gateway National Recreation Area. Final General Management Plan/Environmental Impact Statement April. Accessed at: http://parkplanning.nps.gov/document.cfm?parkID=237&projectID=16091&documentID=59051.

Pahl-Wostl, C. 2009. A conceptual framework for analysing adaptive capacity and multi-level learning processes in resource governance regimes. *Global Environmental Change* 19: 354–365.

Peterson, J.T., and Evans, J.W. 2003. Quantitative decision analysis for sport fisheries management. *Fisheries* 28(1): 10–21.

Peterson, G., Allen, C.R., and Holling, C.S. 1998. Ecological resilience, biodiversity, and scale. *Ecosystems* 1: 6–18.

Peterson, G.D., Cumming, G.S., and Carpenter, S.R. 2003. Scenario planning: A tool for conservation in an uncertain world. *Conservation Biology* 17(2): 358–366.

Pielke, R.A., Jr. 2007. *The Honest Broker: Making Sense of Science in Policy and Politics*. Cambridge, U.K.: Cambridge University Press.

Polasky, S., Carpenter, S.R., Folke, C., and Keeler, B. 2011a. Decision-making under great uncertainty: Environmental management in an era of global change. *Trends in Ecology & Evolution* 26(8): 398–404.

Polasky, S., Nelson, E., and Camm, J. 2008. Where to put things: Spatial land management to sustain biodiversity and economic returns. *Biological Conservation* 141: 1505–1524.

Polasky, S., de Zeeuw, A., and Wagener, F. 2011b. Optimal management with potential regime shifts. *Journal of Environmental Economics and Management* 62: 229–240.

Possingham, H., and Biggs, D. 2012. Resilience thinking versus decision theory? *Decision Point* 62: 4–5.

Regan, H.M., Ben-Haim, Y., Langford, B., Wilson, W.G., Lundberg, P., et al. 2005. Robust decision-making under severe uncertainty for conservation management. *Ecological Applications* 15(4): 1471–1477.

Robinson, K.F., and Fuller, A.K. in press. Guiding stakeholders through the process: Participatory modeling and structured decision making. In: *Including Stakeholders in Environmental Modeling: Considerations, Methods, and Applications*. Gray, S., Paolisso, M., and Jordan, R. (eds). Switzerland: Springer International Publishing.

Robinson, K.F., and Jennings, C.A. 2012. Maximizing age-0 spot export from a South Carolina estuary: An evaluation of coastal impoundment management alternatives via structured decision making. *Marine Coastal Fisheries* 4: 156–172.

Runge, M.C. 2011. An introduction to adaptive management for threatened and endangered species. *Journal of Fish and Wildlife Management* 2: 220–233.

Runge, M.C., Converse, S.J., and Lyons, J.E. 2011. Which uncertainty? Using expert elicitation and expected value of information to design an adaptive program. *Biological Conservation* 144: 1214–1223.

Scheffer, M. 2009. *Critical Transitions in Nature and Society*. Princeton, NJ: Princeton University Press.

Stedman, R.C. 2015. The reification trap, or "following the data around": Resilience and the sustainability hangover. Paper presented at Resilience 2014: Resilience and Development.

Stedman, R.C., Patriquin, M.N., and Parkins, J.R. 2012. Dependence, diversity, and the well-being of rural community: Building on the Freudenburg legacy. *Journal of Environmental Studies and Sciences* 2(1): 28–38.

Tyre, A.J., Peterson, J.T., Converse, S.J., Bogich, T., Miller, D., et al. 2011. Adaptive management of bull trout populations in the Lemhi Basin. *Journal of Fish and Wildlife Management* 2: 262–281.

Waldman, J. 2008. Research opportunities in the natural and social sciences at the Jamaica Bay Unit of Gateway National Recreation Area. Brooklyn, NY: National Park Service. p 77.

Walker, B. 1992. Biodiversity and ecological redundancy. *Conservation Biology* 6: 18–23.

Walker, B. 1995. Conserving biological diversity through ecosystem resilience. *Conservation Biology* 9: 747–752.

Walker, B., and Salt, D. 2006. *Resilience Thinking: Sustaining Ecosystems and People in a Changing World*. Washington, D.C.: Island Press.

Walters, C.J. 1986. *Adaptive Management of Renewable Resources*. New York, NY: MacMillan Publishing Co.

Williams, B.K. 1996. Adaptive optimization of renewable natural resources: Solution algorithms and a computer program. *Ecological Modelling* 93: 101–111.

Williams, B.K. 2001. Uncertainty, learning, and the optimal management of wildlife. *Environmental and Ecological Statistics* 8: 269–288.

Williams, B.K. 2009. Markov decision processes in natural resources management: Observability and uncertainty. *Ecological Modelling* 220: 830–840.

Williams, B.K., Eaton, M.J., and Breininger, D.R. 2011. Adaptive resource management and the value of information. *Ecological Modelling* 222: 3429–3436.

Williams, B.K., and Johnson, F.A. 1995. Adaptive management and the regulation of waterfowl harvests. *Wildlife Society Bulletin* 23: 430–436.

PART IV

Prospects for Resilience in Jamaica Bay

11

Strategies for Community Resilience Practice for the Jamaica Bay Watershed

Laxmi Ramasubramanian, Mike Menser, Erin Rieser,
Leah Feder, Racquel Forrester, Robin Leichenko,
Shorna Allred, Gretchen Ferenz, Mia Brezin,
Jennifer Bolstad, Walter Meyer, and Keith Tidball

The interconnections between community resilience and vulnerability are complex and fraught. They are also different in kind and scope from aspects of ecological resilience described in earlier chapters. Building on interviews conducted with community leaders and experts in 2014 and reported in chapter 6, this chapter highlights best practices that can be adapted to develop the resilience capacity of communities in Jamaica Bay and frames community resilience through a socio-ecological lens (see also chapters 1, 3).

These best practices are informed by analyzing observations made by Jamaica Bay residents after Hurricane Sandy, combined with insights emerging from the literature on social resilience. These two perspectives lead us to recommend strategies for enhancing community resilience targeted at the post-Sandy Jamaica Bay communities (figure 11-1); however, they are likely to have more general relevance for resilience practitioners working in urban estuaries elsewhere as well.

Community resilience and vulnerability can be considered at many scales, including the individual, household, and neighborhood (Rockefeller Foundation, 2014; Jha et al., 2013). Miller et al. (2010) stated that vulnerability encompasses "characteristics of exposure, susceptibility, and coping capacity, shaped by dynamic historical processes, differential entitlements, political economy, and power relations, rather than as a direct outcome of a perturbation or stress." From this perspective, individuals and communities

Figure 11-1. Baby's Dream, Sheepshead Bay, Brooklyn. Many felt their dreams were washed away after Hurricane Sandy. Resilient communities have the ability to bounce back after disasters. This baby gear and clothing store has since reopened. Photograph by Vic Peters on October 31, 2012, courtesy of Wikimedia Commons.

with limited material resources are often more vulnerable to natural disasters and other extreme events because structural factors impair or impede the development of the capacities crucial for resilience. For example, the poor are more likely to live in housing that is structurally suspect (e.g., substandard wiring, leaky roofs and windows) or less well maintained and thus less resilient to many sorts of disasters. In such cases, not only are the poor vulnerable from the point of view of resilience, they suffer in everyday life as well.

Poorer people are especially vulnerable because their ability to bounce back or be resilient is so precarious. They have fewer resources to fall back upon. But vulnerabilities and deficiencies in resilience are present across economic classes. Unemployment, foreclosure, disability, loss of health insurance, collapse of a social network, aging—all of these can befall people regardless of economic class and lead to a significant increase in vulnerability, especially if infrastructure fails or support institutions are inadequate at a moment of crisis. After Hurricane Sandy, some who were not poor before the event were made less secure in a financial and psychological sense as a result of losing their means of earning a livelihood or because they lost their usual place of residence, as reported in chapter 6.

Yet it is also true that although Hurricane Sandy was traumatic for many, if not all, residents around Jamaica Bay, it was also an opportunity for some communities to pull together. As Gotham and Campanella concluded after their 2011 study of the aftermath of Hurricane Katrina in New Orleans, cities do not appear to be either inherently vulnerable or inherently resilient, but rather should be thought of as having elements of each. Creating networks and cross linkages through strategic interventions can help increase the resilience of communities next time a disturbance strikes.

In this chapter we organize our discussion around three themes that emerged from our interviews and focus groups that help shed light on how people living around Jamaica Bay think about the resilience of their communities and landscape. These themes include different ways of knowing and experiencing Jamaica Bay; the role of uneven development in shaping relationships with the bay; and knowledge of Jamaica Bay and perceptions of community resilience. These themes highlight the interactions (or lack thereof) between community residents and the biophysical environment. We build on these themes in the second half of the chapter to suggest strategic ways that different actors can enhance community resilience in the future through education, engagement, communications, self-advocacy, and governmental action.

Different Ways of Knowing and Experiencing Jamaica Bay

The idea of separation from New York City, even though they live within the political boundaries of the city, was repeatedly expressed in interviews with residents. People feel and think of themselves as far removed from the center of "the city." The notion of isolation is a contradictory fact of life for many residents living in and around Jamaica Bay. For many residents who chose to move to the area in search of "island communities," the geographic isolation of Jamaica Bay and its diverse ecologies is a valuable asset and aspect of daily life. The sense of being "baykeepers" or stewards of the coastal marshland and its ecology is often directly correlated to Jamaica Bay's geographic distance and environmental distinctiveness compared with the rest of New York. Residents of Broad Channel and Breezy Point, in particular, emphasize that their lives are in many ways organized around the fluctuations of Jamaica Bay. "Our calendars in Broad Channel have the tides on it," one Broad Channel resident noted. "We live by the tide. For example, tonight I cannot park my car in front of my house." Likewise, in both interviews with local community leaders and focus groups, residents often pointed out how the inlets and shores around Jamaica Bay are perceived to be teeming with fish, birds, and other estuary life.

For many participants, knowledge of the biodiversity and ecosystems of Jamaica Bay came from growing up in the immediate area and experiencing Jamaica Bay as a regular playground or weekend and summer vacation destination. In Broad Channel, in particular, numerous bungalows were owned by multiple families who used them as second homes. Others who grew up in the area also noted learning about Jamaica Bay through more formal institutions, such as secondary schools that had marine biology programs, and the American Littoral Society, which hosts regular educational walks and boating trips around Jamaica Bay. These communities reflected a long-standing relationship to Jamaica Bay, evidenced by intimate understanding of the tides, recreational fishing, bird-watching, concern for environmental protection, and dealing with fairly regular flooding from both Jamaica Bay and high water tables.

In contrast, other residents do not know where Jamaica Bay—the bay itself—is located geographically, or that the bay is distinct from the Atlantic Ocean. Some of these people know the marine waters only in terms of its public access points, such as Canarsie Pier. More significantly, some residents do not understand Jamaica Bay as the complex and dynamic ecosystem that it is to the scientists, experts, and other stewards of the bay. These differences in the general public's understanding of the geography of Jamaica Bay are somewhat correlated with proximity to and frequency of contact with the water's edge and water-based activities. Even in places such as Gerritsen Beach and Breezy Point, communities that have the most direct access (semiprivate/private beaches) to Jamaica Bay, the physical and psychological connection to the bay did not prepare residents to anticipate or plan for the bay to become a threat to their homes and communities.

Members of focus groups conducted in Canarsie, for example, noted that although there was a certain knowledge of proximity to Jamaica Bay in the community, it was in the context of knowing the location of specific landside access points such as Canarsie Pier. But even these understandings may be tenuous. One resident described walking to and from fishing areas in Canarsie with his son, carrying fishing poles on their shoulders, and being stopped by fellow Canarsie residents who incredulously asked, "There's somewhere to fish around here?" Being geographically close to the bay does not automatically translate to a connection or engagement with the bay. Interviewees in some neighborhoods such as Idlewild in Queens, just east of John F. Kennedy International Airport, echoed an often-repeated sentiment that many residents did not know the bay existed, even if the water was only a few blocks away from their homes. In turning their backs to the bay (by choice or circumstance), these residents became vulnerable to the impacts of Hurricane Sandy when floodwaters entered homes and businesses, creating many short- and long-term challenges.

Participants in our Canarsie focus groups had positive feelings about their place of residence. The suburban (less dense) character of the neighborhood, housing affordability, tree-lined streets, parks, and medium-rise (rather than high-rise) developments allowed residents to feel as if they do not live in "the city." Residents in this community did not, however, indicate that proximity to the water had been a motivation for moving to the area. Recent improvements to Canarsie Park and the renovation of Canarsie Pier have encouraged some residents to venture out to Jamaica Bay more regularly. Greenspaces in this neighborhood have also been the sites of community events, such as Canarsie Day, which brings together local community businesses and organizations with the goal of building community alliances and resilience.

The Role of Uneven Development in Shaping Relationships with the Bay

Discrepancies in relationships to Jamaica Bay can, in many ways, be traced to the history of New York's midcentury urban development and planning (discussed in chapters 4–6), which spatially severed numerous Jamaica Bay communities from Jamaica Bay. Projects such as the Belt Parkway, JFK Airport (Idlewild), and a number of city landfills were built essentially between residential communities and Jamaica Bay (Van Hooreweghe, 2012).

Residents and community stakeholders from neighborhoods in proximity to these projects understood and cited this specific history as contributing to why residents in their neighborhoods either have no relationship to Jamaica Bay or view it as "dump." It should be noted that, although none of these landfills are in operation, popular characterizations of Jamaica Bay continue to frequently associate the water (and sometimes neighboring communities) with the landfills and waste. During focus groups in Rockaway Park, participants noted that if they go to the bay they often do so to go "fishing for garbage," explaining that they were more likely to find abandoned materials that they sometimes could salvage rather than fishing as it is conventionally understood. According to the residents, the city government is responsible for many abandoned lots that have been used as dumping sites, either formally or informally (we have not confirmed or denied this assertion). They believe that eventually, any dumping in the area results in pollution of Jamaica Bay and, implicitly, their homes. For residents, the presence of abandoned vacant lots, the persistent challenges of illegal dumping, and the general neglect of the bay convey the impression that their neighborhoods have been deemed unworthy of investment and, furthermore, are places where pollution and waste can accumulate without care. However, some neighborhoods, such as Gerritsen Beach and Broad Channel, have made a conscious choice to protect and maintain public access and use of the bay's shores and waters. These residents often recalled reporting illegal dumping on both city and private property in their own and adjoining neighborhoods.

The distinctions between various aspects of knowing and interacting with Jamaica Bay reflects, to a large extent, how economic and cultural history plays a role in constituting relationships to Jamaica Bay. People who chose to live in communities near Jamaica Bay for the explicit purpose of living near the water or in a beach or island community have a deeper and more amicable relationship to the bay, as opposed to those who live in the area because of economic affordability and access to public housing. A disproportionate amount of New York City Housing Authority housing was built on the Rockaway Peninsula during midcentury urban renewal projects (Van Hooreweghe, 2012). Differences in economic status tend to translate, via reasons for living near the bay, into unequal access and disparate interests in the natural aspects of Jamaica Bay (i.e., Kornblum and Van Hooreweghe, 2010).

Knowledge of Jamaica Bay and Perceptions of Community Resilience

In our interviews, knowledge of and relationships to Jamaica Bay were not often at the forefront of what helps maintain community resilience. Instead the idea of trust and community members taking care of each other characterized community resilience. In some cases, it was clear that shared sentiments about the intrinsic value of nature and of the water helped to organize folks in communities such as Breezy Point, Broad Channel, and Gerritsen Beach around collective actions to maintain the health of the waters and shorelines around them. In other cases, the bay was described as a threat, especially for those community members who had their first encounters with tidal and storm surge flooding during Hurricane Sandy. For these people, understanding, engaging with, and maintaining the ecological health of the bay was not largely considered to be directly connected to the resilience of Jamaica Bay communities.

Several interview participants suggested global climate change and sea level rise were imminent threats to their particular community and other folks around Jamaica Bay (figure 11-2). Yet, many did not feel that global climate change was a regular topic of conversation within their communities, nor was it incorporated in plans to rebuild housing and infrastructure in the wake of Hurricane Sandy. Interviewees noted that rebuilding plans sometimes sought to rebuild homes and communities as they were before the hurricane, not taking into account the possibility of a similar event happening again anytime soon. The idea of Hurricane Sandy as a "once in a lifetime storm" in many neighborhoods seems to put Jamaica Bay in the background of resident conceptions of community resilience.

Building Community Resilience Capacity in Jamaica Bay

In the second half of the chapter, we outline some key strategies to enhance community resilience around Jamaica Bay going forward. These recommendations draw from the literature review described in chapter 3 and the interview results described in chapter 6.

Education

There is a strong need for more education about Jamaica Bay and around the concept of resilience. We recommend that organizations and government-supported groups that are already in place to serve the communities surrounding Jamaica Bay familiarize and

Figure 11-2. Jamaica Bay, looking west toward Manhattan. Resilience begins with how communities conceive of their relationship to the water, land, and city in urban watersheds such as that of Jamaica Bay. Courtesy of Brooklyn College.

educate themselves more thoroughly about the people and communities they serve. This effort will expedite the process of cultivating community resilience by creating an institutional framework that can serve local communities and facilitate the processes of helping them to confront moments of disruption—whether that be ecological, economic, infrastructural, or otherwise. Although individuals within the community certainly can enhance their own understanding of resilience, as described below, such knowledge will prove to be ineffective if local grassroots and government leadership is not in place to facilitate and serve their communities.

Following suit, communities around Jamaica Bay would benefit from learning opportunities such as preparedness workshops and exhibits whereby residents gain more intimate knowledge of Jamaica Bay and common causes of disruption, in particular flooding and climate change. Although some communities have strong local knowledge of the bay, others do not, and this discrepancy is an educational opportunity. Classes in low-income areas that encourage residents to be engaged with Jamaica Bay, including well-advertised and free or low-cost swim lessons, and targeted opportunities to use the parks and the beaches, could serve this end by fostering a sense of place that could translate to a deeper understanding of extreme weather threats.

By developing local knowledge of Jamaica Bay and threats to local communities, residents become positioned to develop a disaster plan. Because so many communities were affected by Hurricane Sandy, this work is already under way; many community members who seek leadership in this arena have plugged into relief and recovery structures such as the long-term recovery groups. However, much of the staffing and effort provided by external nonprofit or short-term grant funding has expired as time has elapsed since Hurricane Sandy. To develop plans and processes that can persist in the long run and reach those community members who are not already engaged, consistent funding will need to be dedicated to staffers who can execute this work.

Engagement

Resilience is derived in part from community engagement and connections. There are significant efforts under way to activate and engage youth and teens in the communities and build bridges to the local environment. More funding and technical support for engagement projects would benefit these efforts, while activating connections to more well-resourced institutions, such as the academic partners within the Science and Resilience Institute at Jamaica Bay consortium and government entities.

Among residents of all ages, more job training and support could provide significant benefit to individuals, families, and communities. More training spaces and opportunities would activate the un- and underemployed population. Local community spaces, although not always suitable for large-scale training, might be suitable for Occupational Safety and Health Administration meetings or pre-apprenticeship trainings that might equip potential employees with soft skills such as making a budget and picking up a paycheck. Drawing people in so that resilience-oriented institutions can get to know them is the first step in a broader conversation.

Communication

Increased communications capacity could significantly augment the resilience of Jamaica Bay communities. We consistently heard focus group participants comment on the lack of communication between various organizations (both local and governmental), leading to an unnecessary duplication of effort, which is particularly frustrating because it wastes groups' already limited resources and impedes efforts to efficiently communicate with the public. Meanwhile, other community needs were described as "slipping through the cracks."

Increasing communication capacity would involve developing the resources to disseminate information through already developed channels, while also building new channels where they do not currently exist. For the segment of the population that looks for information online, news and events should be posted on websites in standardized formats to allow for easy aggregation and curation. It is also important that social media, such as Facebook and Twitter, are used to provide up-to-date information about news and events. Developing organizational capacity to post to the Internet, either on social media or via simple websites, should be a high priority. For those who do not use the Internet frequently or do not have access, news, information, and events should also be published in newspapers and on flyers posted prominently in spaces where residents currently gather. To that end, community-based organizations and organized advocacy groups should develop both physical and virtual places for community engagement or officially sanction and publicize existing physical and electronic locations that provide opportunities for the general public to gather, disseminate, and retrieve information. This "front desk for the community" would significantly augment the resilience capacity of all the neighborhoods surrounding the bay. An increase in the capacity of particular spaces to be resources in general—for offering trainings, connections to social services/government, community knowledge, etc.—can lead to greater potential for those spaces to be used as hubs during times of acute vulnerability. Extra steps should be taken to communicate with and reach the more isolated and vulnerable residents, especially the elderly, the handicapped, and those with language barriers. Organizing efforts toward knowing one's neighbors, establishing block groups, and maintaining emergency call lists would help reach disengaged residents.

Self-Advocacy

Many residents have the capacity and are willing to "self-advocate," and yet the channels to do so remain convoluted and mired in bureaucracy that often outpaces the time and energy of residents. Hurricane Sandy provided a rapid-deployment education and community-organizing opportunity for residents around the bay. It became immediately imperative for community members to develop an understanding of the governmental and nonprofit agencies' programs with jurisdiction over their areas. Unfortunately, the programs often confuse and frustrate residents because of the variety of federal, state, and city programs; confusing language; and bureaucratic tangles. Programmatic failures leave community members uncertain as to how to self-advocate in the face of an uncertain future.

Local capacity and relationship building is a vital step toward developing knowledge about how to interact with government agencies and nonprofits, develop relationships with others attempting to do the same, and enhance the community's ability to coordinate response and take action on issues. Pivoting government-led processes into community-maintained organizing coalitions and networks by providing persisting funding and technical support could contribute significantly toward stronger local capacity and better partnerships. Working with organizations (local and governmental) to help produce better community–institution interfaces would also help to repair a fledgling system and build its capacity for serving and enhancing the lives of the communities they attempt to serve.

Some of the challenges to enhancing community resilience capacity are due to the scope and complexity of Jamaica Bay: it encompasses one hundred miles of coastline, and the bay itself is a dynamic, ever evolving system with extensive littoral drift and an outlying barrier island (as described in chapter 4). Within New York City and the region, the communities around Jamaica Bay are isolated in various ways, from mass transit to jobs and health care. A particular challenge is that many vulnerable people are assigned to the Rockaways by the city's social service system, especially because of the enormous amount of housing dedicated there for low-income, handicapped, and seniors (Alliance for a Just Rebuilding, 2014; New York City-Special Initiative on Recovery and Resiliency, 2013). Thus, there is an overrepresentation of the most socially vulnerable in one of the most ecologically and infrastructurally vulnerable places in the region. Another problem is that in many parts of the Jamaica Bay area, especially the Rockaways, there is not much of a job market, and it may not make sense to create more infrastructure for one (however, low-intensity commercial activity such as restaurants and recreation is desired by the community). This is in contrast to other parts of the low-lying vulnerable parts of the city where there are strong job markets and often decent-paying jobs–for example, in industrial zones such as Hunts Point (Bronx) and Red Hook (Brooklyn). In those places, stronger arguments can be made for more capital-intensive infrastructure resilience.

The Role of Government

As discussed above, affordable housing, living wage jobs, and competent social services are often factors that enhance community resilience capacity. Community resilience capacity also requires that government actively support empowered community participation for several reasons: (1) to better understand the needs of the community; (2) to obtain knowledge from the community about the systems or hazards in question; (3) to educate the community to effectively implement sustainability-resilience programs or proliferate such practices (of stewardship, mitigation, harm reduction, adaptation, etc.); (4) to promote the capacity for community self-organization using law, funding, service delivery, or the formation of social-public partnerships; and (5) to satisfy other ethical, political, or cultural values that are distinct from sustainability-resilience by itself (e.g., social justice, human rights, economic development).

With the possible exception of point 4 (above), most views of sustainability-resilience pursued by existing institutions include these needs and values, though some stress more than others. For example, New York City's sustainability plan, OneNYC, and the Special Initiative for Rebuilding and Resiliency stressed understanding community needs (point

1) and multiple values (point 5). It is instructive to note that there is a shift in the literatures, from sustainability to resilience. Many views of sustainability didn't put much emphasis on the idea expressed in point 4, community self-organization, whereas this idea is a common emphasis in resilience studies. The reason is clear: resilience requires response in an emergency, and basically no government agency has the ability to do this on its own. Responding in the moment requires the knowledge, initiative, and action of everyday people, not just the expert and the official (Camponeschi, 2013; Scruggs, 2014). Indeed, it was common to hear from the victims of Hurricane Sandy around Jamaica Bay that in many neighborhoods the "first responders" were not the police or fire department or emergency medical technicians, but the community itself, sometimes neighbors, sometimes strangers, who wanted to lend a helping hand (Alliance for a Just Rebuilding, 2014).

Acknowledgments: The authors acknowledge Monica Barra, Victoria Curtis, Amanda Lewis, Tim Viltz, and Jeremy Wells for contributing to the field data collection and analyses that informed the strategies suggested in this chapter.

References

Alliance for a Just Rebuilding. 2014. Weathering the Storm: Rebuilding a More Resilient NYC Housing Authority after Hurricane Sandy. New York, NY. Available at: www.rebuildajustny.org.

Camponeschi, C. 2013. Enabling Resilience: Social Innovations in Enabling Sustainability and Participatory Governance, Presentation at 2013 Resilient Cities Conference, Bonn, Germany. Accessed at: http://resilient-cities.iclei.org/fileadmin/sites/resilient-cities/files/Resilient_Cities_2013/Presentations/H3_Camponeschi_RC2013.pdf.

Gotham, K.F., and Campanella, R. 2011. Coupled vulnerability and resilience: The dynamics of cross-scale interactions in post-Katrina New Orleans. *Ecology and Society* 16(3): 12. http://dx.doi.org/10.5751/ES-04292-160312.

Jha, A.K., Miner, T.W., and Stanton-Geddes, Z. (eds). 2013. *Building Urban Resilience: Principles, Tools, and Practice*. Washington, D.C.: International Bank for Reconstruction and Development/The World Bank.

Kornblum, W., and Van Hooreweghe, K. 2010. Jamaica Bay Ethnographic Overview and Assessment. Northeast Region Ethnography Program, National Park Service, Boston, MA.

Miller, F., Osbahr, H., Boyd, E., Thomalla, F., Bharwani, S., et al. 2010. Resilience and vulnerability: Complementary or conflicting concepts? *Ecology and Society* 15(3): 11. Accessed at: http://www.ecologyandsociety.org/vol15/iss3/art11/.

New York City-Special Initiative on Recovery and Resiliency. 2013. Chapter 16: South Queens. In: *PlaNYC: A Stronger, More Resilient New York*. Accessed at: http://www.nyc.gov/html/sirr/downloads/pdf/final_report/Ch16_SouthQueens_FINAL_singles.pdf.

Rockefeller Foundation. 2014. The City Resilience Framework. April. Accessed at: https://assets.rockefellerfoundation.org/app/uploads/20140410162455/City-Resilience-Framework-2015.pdf.

Scruggs, G. 2014. Participatory budgeting's birthplace uses the mechanism to build resilience. Next City. July 3. Accessed at: https://nextcity.org/daily/entry/participatory-budgetings-birthplace-uses-the-mechanism-to-build-resilience.

Van Hooreweghe, K.L. 2012. The Creeks, Beaches, and Bay of the Jamaica Bay Estuary: The Importance of Place in Cultivating Relationships to Nature. PhD dissertation. City University of New York.

12

The Future of Jamaica Bay: Putting Resilience into Practice

Adam S. Parris, William D. Solecki, Eric W. Sanderson, and John R. Waldman

Hurricane Sandy shook New York City to its core. In the immediate aftermath, the emphasis was on recovery, but in the long-term perspective, Hurricane Sandy motivated the city as a whole to think about how it can not only be a sustainable place to live, but also be resilient to different kinds of shocks. This long-term view brought Jamaica Bay to the forefront of resilience discussions. The waters, wetlands, and communities of the Jamaica Bay watershed had long been a focal point for revitalizing a heavily stressed and affected coastal area through habitat restoration, improvements to public access, outdoor recreation, and sustainable development. City, state, and federal governments, all with jurisdiction in the bay, had sought science-based solutions to achieving integrated coastal zone management, with various levels of success. They also recognized that old approaches were not entirely working. Against this backdrop, the effects of Sandy on Jamaica Bay and the surrounding communities dramatically increased the urgency for putting new solutions into action.

It is fortuitous that even before the hurricane, the City of New York and the National Park Service had issued a call for a consortium of institutions to respond to the long-term drivers of change around Jamaica Bay. The Science and Resilience Institute at Jamaica Bay (hereafter, the institute) (www.srijb.org) arose from that call to action. The institute produces integrated knowledge to increase biodiversity, human well-being, and adaptive capacity in communities and waters surrounding Jamaica Bay and New York City. In the process, it advances innovative thinking and learning about the resilience of urban coastal regions through programs of research and engagement. The institute, hosted by Brooklyn College, is a partnership among academic institutions, government agencies, nongovernmental organizations, and community groups. Core partnerships are sustained among the National Park Service, the City of New York, and a consortium that includes the City University of New York, Columbia University, Cornell University, Rutgers University's Institute of Marine and Coastal Sciences, NASA Goddard Institute for Space Studies, New York Sea Grant, Stevens Institute of Technology, Stony Brook University (SUNY), and the Wildlife Conservation Society. This volume is one of the first products of the institute's research, hence its forward-looking focus on "prospects" for resilience. We realize we are starting at the beginning: we are not relying on one specific ten-point plan or one specific resilience framework. As the previous eleven chapters indicate, instead we have diverse ideas, a wealth of information, and possible solutions. We have beginnings, not ends. Our struggle is not unique. Others working in urbanized watersheds in other parts of the world may be having similar issues. We have written this book to be a foundation not only for our efforts but also for resilience elsewhere.

This concluding chapter illustrates ways in which the institute functions at the interface of science, policy, and practice. Specifically, the institute facilitates communication and dialogue among different kinds of people and institutions through informal and formal engagement. By doing so, the institute fosters knowledge exchange and the coproduction of knowledge for actions that promote resilience, similar to a "boundary organization" (Guston, 2001; Cash et al., 2003). We begin by discussing how science, policy, and practice are linked in the context of resilience, then draw on this background and all the chapters in the volume to develop shared observations about the "prospects" for resilience and the future of Jamaica Bay.

Science, Policy, and Practice in the Context of Resilience

Congress recognizes the coastal zone as an important area for a variety of complex uses, from commerce and development to ecosystem diversity to recreation (Coastal Zone Management Act, 16 U.S.C. §§1451–1465). These uses are particularly concentrated in urban coastal areas because a great deal of the population and its commercial activity are located in cities. With these complex and conflicting uses, coastal areas are affected by numerous and interrelated stressors, many of which are exacerbated by climate-related hazards (Moser et al., 2014). The impacts affect our ability to ensure clean water for people (Clean Water Act, 33 U.S.C. §§1251–1387) and to protect habitat for endangered species (Endangered Species Act, 16 U.S.C. §§1531–1544). Thus, resilience in urban coastal areas is affected by actions at multiple levels from individuals and community groups to city, state, and federal governments. At any level, actions can be a response to immediate local hazards (e.g., flooding of coastal homes) or long-term drivers of global change (e.g., sea level rise). With its concentrated and complex history of coastal zone management, Jamaica Bay is symbolic of the coastal legacy of the United States and set within the context of a globally relevant city. For these reasons, we argue it is a sentinel site for applying resilience concepts to reshape this legacy.

Science can help inform this process of revitalization in Jamaica Bay in that it educates, expands alternatives, clarifies choices, and aids in formulating and implementing decisions (Sarewitz and Pielke, 2007). Decades of work analyzing the process of using science to support decision making emphasize the importance of having credible scientific information with relevance or salience to the problem and legitimate processes of research without political suasion or bias (Cash et al., 2003; McNie et al., 2016). The institute's research consortium has a long track record of producing credible, if somewhat disjointed, science on Jamaica Bay, but an important future role is to increase the salience and legitimacy of that science, particularly for resilience-focused research. One way to facilitate salience and legitimacy is to give public agencies and community stakeholders a larger role in shaping scientific research agendas (Dilling et al., 2015; Dow et al., 2013; NRC, 2009; Owen et al., 2012; Simpson et al., 2016).

The institute is, by definition, a partnership that facilitates this role. The City of New York, the National Park Service, and the research consortium have a general agreement that calls for the institute to consider the advice of public agencies and stakeholders in the design and prioritization of research. To this end, the institute convenes two advisory committees—a public agency committee and a stakeholder advisory committee. Each group advises the institute on their respective views on science and knowledge gaps. They enable the institute to establish a collaborative, cutting-edge science framework.

Stemming from these committees, the institute aims to build a wider network of research partnerships among three audiences: government, science, and local communities. Across these three audiences, the institute looks to link "communities of practice"— people who care about a shared set of issues or approach them with a common set of ideas and techniques—in periodic and iterative resilience-oriented discussion (figure 12-1). The institute has organized charrette-type workshops with both members of committees and scientific modelers to identify new design concepts that can be modeled to assess resilience. The network of participants involved in institute activities includes people in other cities, estuaries, and resilience-oriented programs around the United States and abroad (figure 12-1).

As scientists engage Jamaica Bay decision makers and stakeholders, they must consider not only whether science is used to inform decisions, but how it is being used. In highly political contexts or where science is used for decisions of great consequence, there is a need for attention to boundary management (McNie et al., 2016; Cash et al., 2003; Guston, 2001; Sarewitz, 2004). Boundary management involves communicating between science and society, translating information, and mediating and negotiating across the boundary. Boundary management is not typically associated with an organization whose

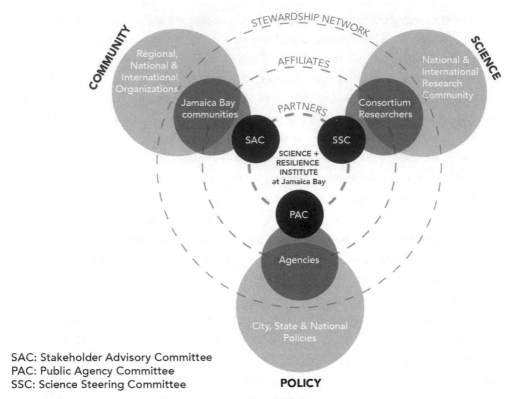

SAC: Stakeholder Advisory Committee
PAC: Public Agency Committee
SSC: Science Steering Committee

Figure 12-1. Governance and partnership framework for the Science and Resilience Institute at Jamaica Bay (SRIJB). Through formal partnerships with public agencies and Jamaica Bay communities, the institute is better able to connect resilience-focused research to decision making and stewardship of Jamaica Bay. The institute also provides a framework for integrating different forms of scientific knowledge stemming from resilience thinking; reconciling that knowledge with what we know from policy and practice; and managing interactions for healthy, ongoing dialogue geared toward learning. Figure courtesy of Jessica Fain, SRIJB.

primary mission is research and knowledge generation, such as a university. Nor is it the role of management agencies, who have regulated and therefore, constrained, functions within government. Resilience practice—given the complexity of different and competing values, multiple jurisdictions, and diverse communities explained in the earlier chapters of this book—needs an organization working at the edges of scientific disciplines, across jurisdictions, and among the many public and management agencies (highlighted in chapter 1). In Jamaica Bay, and in other urban watersheds, the coproduction of knowledge and the coordination of resilience practice are a promising endeavor to realize new solutions (Meadow et al., 2015). In this regard, the institute is a multifaceted organization whose mission reflects not only the prospects of resilience-focused research but also the prospects of greater institutional capacity at the interface of science, policy, and practice.

Prospects for Resilience

This book is the first sustained attempt by the institute to coordinate and drive those kinds of cogenerative processes, with the goal of learning through the process what the prospects for resilience of the Jamaica Bay watershed really are. We started out with the concept of an assessment report to get down on paper what we knew about resilience of Jamaica Bay. As the sections slowly developed, we realized gaps in what we were discussing, so we sought out additional scientists and scholars to bring in new perspectives. Over time the report morphed into the book you are holding now.

Here we summarize some of the observations that emerged from the process of creating *Prospects for Resilience*, with the goal of informing our own work as part of the institute, and helping other boundary organizations and individuals in other watersheds learn from our experience so that they can make their own urban estuaries more resilient.

The first and most obvious point is that resilience practice is and likely will always be in a state of becoming: it is as much a process as a product. One finds this theme repeatedly throughout the book in different contexts, from the discussion of alternative resilience frameworks in chapters 2 and 3, to the exploration of models in chapter 8 and decision support science in chapter 10, through to the discussion of community resilience in chapter 11. The establishment of the institute and the harrowing experience of Hurricane Sandy (chapter 1) is evidence that resilience is broadly adopted in New York City (City of New York, 2013) and in Jamaica Bay, in particular. However, there remains much about resilience as a framework for change that is contested.

Given the magnitude of past, present, and future disturbance and the massive investment in the city as it exists today, we wonder how will transformation occur? Will it happen through deliberate adaptation or will it involve incremental responses to the exigencies of particular shocks and stresses? The frequency and magnitude of coastal flooding is certain to increase as the population of New York City grows, but the exact timing of storm surge remains unpredictable more than a few days out from the storm. Moreover, we recognize that particular disturbances are not the only part of the story. The institute can help the city balance its response to these hazards with its goal to achieve equity and sustainability, particularly in economically depressed areas, many of which huddle around Jamaica Bay. How do freshwater inflows from wastewater treatment plants or the presence of John F. Kennedy International Airport limit or enable options designed to achieve properties of resilience? As we see throughout the book, in many ways the search for prospects for future resilience requires coming to terms with the past drivers of loss in an urbanized watershed such as Jamaica Bay (chapters 4–6).

There is disagreement about which values should be maintained in the light of disturbance around Jamaica Bay. Although most actors in the Jamaica Bay watershed might agree generally that the goals include consideration of a combination of social, economic, and ecological values (chapters 2 and 3), much remains unsaid and unprocessed about how those values are defined, how they are monitored, and how they are prioritized (chapters 10 and 11). Lack of consensus on this question leaves open a fundamental aspect of resilience that requires future work. We struggle to formulate the exact list of values, and we recognize that different voices have different senses of what matters, now and into the future. Here is a role of the institute as a boundary organization. And that boundary work, in itself, can help build resilience. For most of the last four hundred years, trade-offs have consistently gone the way of economic imperatives trumping social and ecological ones, but despite this history it now appears that values are changing in the twenty-first century for city decision makers and in communities around the bay (figure 12-2).

Figure 12-2. Trail through the woods near Jamaica Bay. Resilience is complicated, there is no doubt, but there are also paths of connection through the tangle of issues. Organizations such as the institute discussed in this chapter create the pathways to resilience by working across boundaries between science, policy, and practice. Courtesy of the National Park Service.

These considerations no doubt emphasize the necessity of developing better integration among the social, ecological, and technological (infrastructure) elements on the ground (as discussed in chapter 9) and in our conceptual infrastructure for the bay and its watershed (chapters 4–6). How does Jamaica Bay work as a social-ecological system (SES)? Historical, functional, and management perspectives on Jamaica Bay all point toward the same conclusion: no one component of Jamaica Bay—or any place, for that matter—can be understood fully without also understanding the connection to the other components. Interconnectivity applies when we are discussing the specifics of a wetland restoration project or expansion of the airport. It applies to how we conceive the general and conceptual connection between hydrodynamics and economic development. Some facts and relationships we already know, but other relationships are beyond our grasp. Tools such as monitoring (chapter 7) and modeling (chapter 8), coupled with advancing the basic science of physical (chapter 4), ecological (chapter 5), and social systems (chapter 6), can help us make progress. What is clear is that there are many different resilience knowledge bases and centers of action already present within the people and institutions that surround Jamaica Bay. These centers include academic (expert) knowledge, local community knowledge, and management knowledge. These different forms and sources of data, information, and learning need to be recognized and integrated for effective resiliency efforts (chapters 1, 3, 10, and 11).

In an analogous way, the health and resilience of the waters of the bay depend on the future health and resilience of the surrounding lands in the watershed. The landscape of Jamaica Bay has evolved dramatically over the centuries, and we can only expect future changes. Bulkheads built for a port that never materialized affect the ways that waves are propagated across the bay, in turn affecting the persistence of salt marshes. Paving one's driveway increases stormwater runoff that, when contaminated, leads to more pollution in the bay and changes in the phytoplankton that may or may not affect the fish, birds, and tourists using the bay. One lesson that carries through is that no change occurs without affecting some other aspect of the SES.

Because of these interconnections, we see the need for system-level analysis of Jamaica Bay. Process models can illustrate the components and flows within and around the Jamaica Bay watershed and how the watershed fits into the social-political-ecological regional context. Recent work undertaken by the RAND Corporation is helping to build the beginnings of an integrated modeling framework for the institute (Knopman and Fischbach, RAND, pers. comm.) The RAND–led effort highlights a set of shared values among the management agencies (coastal flood risk reduction, improvements to water quality, and ecosystem health) and suggests how a set of existing models could be wired together to evaluate different scenarios of interaction (chapter 8). Building on that work, we see that future analysis necessarily needs to include the expertise and knowledge on the physical, ecological, and social systems of the bay, recognizing that some of these domains are more advanced than others. Systems analysis, including models, can highlight gaps in understanding and data that will help drive the process of promoting resilience concepts forward.

On a related note, the institute and its partners in the Jamaica Bay watershed need robust resilience indicators and an observational infrastructure to monitor the health of the overall system. The indicators highlighted in chapter 7 can be broadened in scope to include the working concept of ecosystem goods and services, which reflect the physical, ecological, and social values desired for the bay. In this fashion, the resilience of each component part of the system can be clearly connected to a set of driving relationships and for the foundation of long-term monitoring plans. Collecting long-term social and environmental observations is essential for assessing resilience. Otherwise we will not know the status of the values we hope to make resilient, be able to say how they responded when a disturbance occurs, or know whether or when the system has recovered.

Institute scientists are developing participatory tools to visualize and analyze indicators based on ideas about future resilience in Jamaica Bay. Visionmaker is a promising method to analyze metrics and models (chapter 8). It is a platform to create and share visions for Jamaica Bay (and other parts of New York City). Visions constitute combinations of ecosystems, lifestyle choices, and climate scenarios. These are analyzed in terms of underlying system models that produce a set of metrics about how each vision supports the physical, ecological, and social components of New York City. Because Visionmaker is deployed over the Internet (Visionmaker.nyc), it also enables sharing of concepts with a wider audience. Although only in beta release and testing as of this writing, Visionmaker is the kind of tool that a broad set of stakeholders could use collaboratively to test and design resilience interventions for Jamaica Bay. We see it as a tool for ecological democracy.

Boundary tools such as Visionmaker enable different groups of people to work together in the name of resilience, yet allow them to express and work through differences (chapter 11) in a structured way (chapter 10). A wide variety of perceptions of the bay exist (chapters 3 and 6). Some emphasize the ecological values of Jamaica Bay and see it as a place of tourism and recreation. Others see the bay as their home and place of employment. Many agencies have jurisdictional responsibilities and prerogatives for the bay. Others see Jamaica Bay as a threat from flooding that needs to be contained, while others see it as the place where society returns its treated wastewater to nature. Some know little about Jamaica Bay itself except as the reason the Belt Parkway curves or as the surprisingly pleasant green and blue view out a plane window. The recognition and incorporation of different perceptions are important as are slowly and carefully reconciling the assumptions people make about the future. Those tensions are implicit in many of the chapters of this book.

Finally, we conclude, much where we began, with the acknowledgment of the importance of the recent agreement enacted by the City of New York and the National Park Service to try to manage Jamaica Bay jointly as a sustainable and resilient SES. This political decision not only created the basis for the institute, but it also provides the template for shared action in the future. Many of the problems of Jamaica Bay in the past were created by different institutional actors working on their own priorities and not appreciating the long-term significance of their actions. Aligning politics; supporting the interface of science, policy, and practice; documenting what is known; building models that depict different futures—are all essential steps in building what has long been missing for Jamaica Bay—real prospects for resilience in the future (figure 12-3).

Figure 12-3. Ospreys nest in sight of houses and high rises in New York City's Jamaica Bay. SESs can thrive by bringing waters, land, wildlife, and people together in a system of co-dependence and reliance. By living in place, with cognizance of the long-term goals and mechanisms to engage everyone, we can build prospects for resilience. Courtesy of the National Park Service.

References

Cash, D.W., Clark, W.C., Alcock, F., Dickson, N.M., Ecley, N., et al. 2003. Knowledge systems for sustainable development. *Proceedings of National Academy of Sciences* 100: 8086–8091.

City of New York. 2013. A Stronger, More Resilient New York. New York City Special Initiative on Rebuilding and Resiliency. City of New York. Accessed at: http://www.nyc.gov/html/sirr/downloads/pdf/final_report/001SIRR_cover_for_DoITT.pdf.

Clean Water Act of 1972, 33 U.S.C. §§1251 et seq. (2002).

Coastal Zone Management Act. 16 U.S.C. §§1451–1465 (1972).

Dilling, L., Lackstrom, K., Haywood, B., Dow, K., Lemos, M.C., and Berggren, J. 2015. What stakeholder needs tell us about enabling adaptive capacity: The intersection of context and information provision across regions in the United States. *Weather,*

Climate and Society 7: 17. doi:10.1175/WCAS-D-14–00001.1.

Dow, K.D., Haywood, B., Kettle, N., and Lackstrom, K. 2013. The role of ad hoc networks in supporting climate change adaptation: A case study from the Southeastern United States. *Regional Environmental Change* 13: 1235–1244. doi:10.1007/s10113-013-0440-8.

Endangered Species Act. 16 U.S.C. §§1531–1544 (1973).

Guston, D.H. 2001. Boundary organizations in environmental policy and science: An introduction. *Science, Technology, and Human Values* 26: 399–408.

McNie, E., Parris, A., and Sarewitz, D. 2016. Improving the public value of science: A typology to inform discussion, design and implementation of research. *Research Policy* 45: 884–895.

Meadow, A.M., Ferguson, D.B., Guido, Z., Horangic, A., Owen, G., and Wall, T. 2015. Moving toward the deliberate coproduction of climate science knowledge. *Weather, Climate, and Society* 7: 179–191.

Moser, S.C., Davidson, M.A., Kirshen, P., Mulvaney, P., Murley, J.F., et al. 2014. Coastal zone development and ecosystems (Chapter 25). In: *Climate Change Impacts in the United States: The Third National Climate Assessment*. Melillo, J.M., Richmond, T.C., and Yohe, G.W. (eds). U.S. Global Change Research Program. Accessed at: http://www .globalchange.gov/nca3-downloads-materials.

NRC (National Research Council). 2009. *Informing Decisions in a Changing Climate. Panel on Strategies and Methods for Climate-Related Decision Support*. Washington, D.C.: National Academies Press.

Owen, G., McLeod, J.D., Kolden, C.A., Ferguson, D.B., and Brown, T.J. 2012. Wildfire management and forecasting fire potential: The roles of climate information and social networks in the Southwest United States. *Weather, Climate, and Society* 4: 90–102.

Sarewitz, D. 2004. How science makes environmental controversies worse. *Environmental Science & Policy* 7: 385–403.

Sarewitz, D., and Pielke, R.A. 2007. The neglected heart of science policy: Reconciling supply of and demand for science. *Environmental Science & Policy* 10: 5–16.

Science and Resilience Institute at Jamaica Bay. 2016. Accessed at: www.srijb.org.

Simpson, C., Dilling, L., Dow, K., Lackstrom, K.J., Lemos, M.C., and Riley, R. 2016. Assessing needs and decision contexts: RISA approaches to engagement research. In: *Climate in Context: Science and Society Partnering for Adaptation*. Parris, A., Garfin, G., Dow, K., Meyer, R., and Close, S.L. (eds). London: John Wiley & Sons.

Contributor Biographical Sketches

Shorna Allred

Shorna Allred is an associate professor and associate director of the Human Dimensions Research Unit in the Department of Natural Resources at Cornell University. Allred's applied social science research and outreach program is centered on developing an understanding of conservation attitudes and behavior at individual and community levels. Her research has focused on the social dimensions of climate change mitigation and adaption, including community resilience to flooding, riparian landowner decision making, and municipal official motivations and barriers to climate adaptation. Allred holds a PhD from the Department of Forest Ecosystems and Society at Oregon State University (2001).

Jennifer Bolstad

Jennifer Bolstad, RLA, is a principal at Local Office Landscape & Urban Design, LLC. Founded by Harvard Graduate School of Design classmates Bolstad and Walter Meyer, Local Office seeks to ameliorate the impact of cities on the sea, while protecting cities from sea surges. The firm has garnered accolades from across the disciplines of architecture, landscape architecture, public policy, science, and art.

Brett Branco

Brett F. Branco is an assistant professor of earth and environmental sciences and associate director of the Aquatic Research and Environmental Assessment Center at Brooklyn College of the City University of New York. He has served on the executive council of the Science and Resilience Institute at Jamaica Bay (SRIJB) since its inception in 2012 and is a member of the Urban Sustainability Program Steering Committee at Brooklyn College. His research focuses on understanding the dynamics and water quality of shallow urban water bodies, and he's worked in Jamaica Bay and New York City's parks since 2009. Branco holds a PhD in oceanography (University of Connecticut, 2007).

Mia C. Brezin

Mia C. Brezin is the assistant district manager at Manhattan Community Board 11, serving the neighborhood of East Harlem. Brezin holds a master's degree in urban planning (City University of New York [CUNY], Hunter College, 2014).

Monica Bricelj

Monica Bricelj is a senior research professor at the Haskin Research Laboratory, Department of Marine and Coastal Sciences, Rutgers University, and consultant in marine sciences. She has written more than eighty-five peer-reviewed publications on the ecophysiology of bivalve mollusks and their interactions with harmful algae in estuarine coastal waters. She holds a PhD in coastal oceanography from Stony Brook University, New York, where she held a position as associate professor until 1996. Previously she led the shellfish research program for more than ten years at the National Research Council, Halifax, Canada. She is a recipient of a scholarship from the Fulbright Commission in Mexico (COMEXUS) (2015–2016).

Katherine Bunting-Howarth

Katherine Bunting-Howarth is the associate director of New York Sea Grant and the former director of water resources for the State of Delaware. She serves on multiple boards and committees, including the executive council of the SRIJB, New York Water Resource Institute, Great Lakes Basin Advisory Council, Cornell Biological Field Station, and Chesapeake Bay Program Science and Technical Advisory Committee. Bunting-Howarth holds a PhD in marine studies (concentration in marine policy) (University of Delaware, 2001) and a JD with a certificate in environment and natural resource law (University of Oregon, 1995).

Joanna Burger

Joanna Burger is a distinguished professor of biology at Rutgers University. Her major interests are social behavior of vertebrates, ecotoxicology, and stakeholder involvement. She has written more than twenty books and four hundred refereed papers on these topics. Her most recent book, *Habitat, Population Dynamics, and Metal Levels in Colonial Waterbirds: A Food Chain Approach*, with M. Gochfeld, will be published in 2016. Other published books are about animals in urban environments; common terns; black skimmers; and social interaction of seabirds, stakeholders, and scientists; among others. Burger holds a PhD in behavioral ecology from University of Minnesota (1972) and an honorary degree from the University of Alaska, Fairbanks (2006).

Russell L. Burke

Russell L. Burke is a professor of biology at Hofstra University. His research interests focus on the ecology, evolution, and conservation biology of vertebrates. He has conducted long-term research studies on diamondback terrapins at Jamaica Bay. He holds a PhD in biology from the University of Michigan.

Merry Camhi

Merry Camhi is the director of the Wildlife Conservation Society's New York Seascape Program, which seeks to conserve threatened marine wildlife and habitats in the New York Bight through field research, policy initiatives, and public education and stewardship. Camhi holds a PhD in ecology from Rutgers University. As a scientist with the National Audubon Society, she focused on shark and ray conservation, and has served on the IUCN Shark Specialist Group since 1994. Her publications include *The Conservation Status of Pelagic Sharks and Rays* (IUCN, 2009) and *Sharks of the Open Ocean* (Wiley-Blackwell, 2008).

Maria Raquel Catalano de Sousa

Maria Raquel Catalano de Sousa is a research scientist at the National Institute of Industrial Property of Brazil. She holds a BA in civil engineering from Federal University of Santa Catarina, a master's in general and applied hydrology from the Center for Studies and Experimentation of Public Works (Madrid, 2005), a master's in environmental engineering from Federal University of Santa Catarina (Florianopolis, 2006), and a PhD in environmental engineering from Drexel University (Philadelphia, 2015). Her current and recent areas of research include green infrastructure performance under extreme events, urban vegetation responses to climate change, and evaluation of green infrastructure environmental benefits.

Robert Chant

Robert Chant is a professor in the Marine and Coastal Science Department at Rutgers University. An amateur sailor and professional clam digger from Long Island, Chant holds a PhD in oceanography from SUNY Stony Brook and an undergraduate degree in electrical engineering from SUNY Buffalo. His research focuses on the physics of estuarine and coastal systems.

Christina Colón

Christina Colón is assistant professor of biological sciences at Kingsborough Community College. Previously she was a curator at the New York Botanical Garden and research associate for the Wildlife Conservation Society. She conducted her master's thesis at Cockscomb Basin Jaguar Preserve in Belize, and her dissertation on the ecology of the Malay civet (*Viverra tangalunga*) in Borneo. She currently studies the breeding ecology of horseshoe crabs (*Limulus polyphemus*) in Brooklyn. During winter, she radiotracks translocated urban civets in Singapore, and has studied the role of carnivores in rainforest regeneration. Colón holds a PhD in ecology (Fordham University, 1999).

Michael J. Dorsch

Michael J. Dorsch is a doctoral candidate in the Earth and Environmental Sciences Program at the Graduate Center, City University of New York (CUNY). He has served as program support assistant/managing editor of *Prospects for Resilience* at the SRIJB and as a research assistant with the CUNY Institute for Sustainable Cities. His doctoral research is on energy infrastructure transitions and transformations. Dorsch holds a master's in international relations and comparative politics and policy (West Virginia University, 2011).

Bryce DuBois

Bryce DuBois is a post-doctoral associate in the Department of Natural Resources at Cornell University. His dissertation research was on the political ecology of Rockaway Beach, New York City, post–Hurricane Sandy. DuBois holds an MPhil and PhD in environmental psychology (Graduate Center, CUNY, 2013 and 2016, respectively).

Mitchell Eaton

Mitchell Eaton is a research ecologist with the U.S. Department of Interior's Southeast Climate Science Center. He is based at North Carolina State University, where he holds a faculty appointment in the Department of Applied Ecology. Eaton develops models to understand spatial and temporal variation of species and habitats. His research focuses on assisting resource managers with the information they need to make better decisions for conserving trust responsibilities under uncertainty, accounting for other competing societal values. He earned his PhD in ecology and evolutionary biology from the University of Colorado in 2009.

Leah Feder

Leah Feder holds a master's degree in urban planning (Hunter College, 2015).

Gretchen S. Ferenz

Gretchen S. Ferenz is retired senior extension associate and program and resource development specialist, urban environment, Cornell University Cooperative Extension–New York City, and current partner at Tom Fox & Associates. She is a fellow at the Atkinson Center for a Sustainable Future, Cornell University. Ferenz served as founding member of the SRIJB and served on the executive council and committees in support of its creation and growth. Ferenz resides by Jamaica Bay in Breezy Point on the Rockaway Peninsula and is active in the local and regional coastal community. She holds an MS in environmental horticulture from the University of California at Davis, 1984.

Jordan R. Fischbach

Jordan R. Fischbach is a policy researcher at the RAND Corporation and codirector of RAND's Water and Climate Resilience Center. He has expertise in risk analysis, exploratory simulation modeling, and robust decision making, a method designed to better manage deep uncertainty and develop robust and adaptive plans through quantitative scenario analysis. Fischbach works with government agencies to incorporate deep uncertainty into climate adaptation and resilience planning efforts. He earned a PhD in policy analysis from the Pardee RAND Graduate School in 2010, where he was awarded the Herbert Goldhamer Memorial Award.

Racquel Forrester

Racquel Forrester is a workforce and community development manager at the Southwest Brooklyn Industrial Development Corporation and holds a master's degree in urban planning (Hunter College, 2015).

Angela K. Fuller

Angela K. Fuller is the leader of the New York Cooperative Fish and Wildlife Research Unit and associate professor in the Department of Natural Resources at Cornell University. Her research focuses on applied conservation and management of mammals, specifically related to population dynamics and the influence of human-induced landscape changes on populations. Another major program area of her research is applying structured decision making and adaptive management for aiding natural resource management and policy decisions. Fuller is coeditor of the book *Martens and Fishers* (Martes) *in Human-Altered Environments: An International Perspective* (Springer, 2004) and has numerous publications in peer-reviewed journals. Fuller received a PhD in wildlife ecology from the University of Maine (2006).

Mario Giampieri

Mario Giampieri is a program officer at the Wildlife Conservation Society (WCS) and a geography research assistant at Hunter College. He holds a BA in environmental and metropolitan studies (New York University, 2012) and a certificate in geographic information systems (Hunter College, 2015).

Arnold Gordon

Arnold Gordon is a professor of oceanography at the Department of Earth & Environmental Sciences, Columbia University, and is on the research staff at the Lamont-Doherty

Earth Observatory, in Palisades, New York. He also serves on the executive council of the SRIJB. His research is directed at the ocean's stratification and circulation and its linkage to the climate system. He has worked across the global ocean, with focus on the Southern Ocean, the South Atlantic, and the seas of Southeast Asia, as well as more locally in urbanized Jamaica Bay. Gordon holds a PhD in oceanography from Columbia University (1965).

Steven Handel

Steven Handel, distinguished professor of ecology and evolution at Rutgers University, studies plant population ecology, the restoration of urban and degraded habitats, and how these can mesh with landscape architecture design. He has also taught at Yale, Harvard, and Stockholm Universities. He is the editor of the journal *Ecological Restoration*. He received the Theodore Sperry Award in 2011 from the Society for Ecological Restoration for his research on urban habitat creation, and was named an honorary member of the American Society of Landscape Architects. He received his PhD in ecology and evolution from Cornell University in 1976.

Matthew P. Hare

Matthew P. Hare is associate professor of Cornell University's Department of Natural Resources. He holds a PhD from the University of Georgia and was a postdoctoral associate at Harvard University with Dr. Stephen Palumbi. Hare's research is aimed at understanding the ecological, demographic, and historical processes that generate organismal diversity in coastal ecosystems, in addition to understanding the impacts of management practices.

Olaf P. Jensen

Olaf P. Jensen is an assistant professor at the Department of Marine and Coastal Sciences at Rutgers University. He earned a PhD at the University of Wisconsin, Madison, in 2007, and was a David H. Smith conservation research fellow at the University of Washington from 2008 to 2010. His research interests focus on the social-ecological system of fisheries and the ecology of aquatic ecosystems.

Fred A. Johnson

Fred A. Johnson is a research wildlife biologist at the U.S. Geological Survey's Wetland and Aquatic Research Center in Gainesville, Florida. He has authored more than seventy peer-reviewed scientific publications, and has more than twenty years of experience in the application of decision science in natural resource management. He is particularly interested in the need to address the inherent tension between decision analysis and a resilience-based approach to conservation. Johnson holds a PhD in wildlife ecology and conservation (University of Florida, 2010).

Christina M.K. Kaunzinger

Christina M.K. Kaunzinger is senior ecologist at the Center for Urban Restoration Ecology, Rutgers University. Her recent restoration and public education projects include the Rebuild by Design BIG "U" to protect Manhattan from sea level rise and storm surges; restoration and resilience building during infrastructure improvement along the Shore Parkway, Jamaica Bay, New York City; and stewardship and sustainability messaging at Duke Farms, Hillsborough, New Jersey. Current research examines opportunities for inland migration of coastal habitats in response to sea level rise and restoration of American chestnut to northeastern forests. Kaunzinger holds a PhD in ecology & evolution (Rutgers University, 2000).

Debra Knopman

Debra Knopman is a principal researcher at the RAND Corporation and a professor at the Pardee RAND Graduate School. She served as vice president and director of RAND Infrastructure, Safety, and Environment (later Justice, Infrastructure, and Environment) from 2004 to 2014. She holds a PhD in geography and environmental engineering from Johns Hopkins University. She is currently leading an integrated modeling effort for Jamaica Bay.

Jake LaBelle

Jake LaBelle is the research program officer in the Wildlife Conservation Society's New York Seascape Program. He oversees various field studies under way in the New York area, including acoustic and satellite tagging of several shark species in New York waters, as well as monitoring American eels in the Bronx River. LaBelle holds an MA in marine conservation and policy (Stony Brook University, 2012).

Robin Leichenko

Robin Leichenko is professor and chair of geography at Rutgers University and co-director of the Rutgers Climate Institute. Her research explores economic vulnerability to climate change, equity implications of climate adaptation, and the interplay between climate extremes and urban spatial development. Leichenko served as a review editor for the Intergovernmental Panel on Climate Change (IPCC) Fifth Assessment Report. Her book, *Environmental Change and Globalization: Double Exposures* (2008, Oxford University Press), won the Meridian Book Award for Outstanding Scholarly Contribution from the Association of American Geographers.

John Marra

John Marra is a professor in the Department of Earth & Environmental Sciences at Brooklyn College (CUNY), and also director of the college's Aquatic Research and Environmental

Assessment Center. He serves on the executive council of the SRIJB. His research in ocean-ography spans the open ocean, coastal waters, and Jamaica Bay, from phytoplankton to fish. He is the author or coauthor of more than 150 scholarly publications. Marra received a PhD in biological oceanography in 1977 from Dalhousie University in Halifax, Nova Scotia, Canada.

Mike Menser

Mike Menser teaches philosophy and urban sustainability studies at Brooklyn College, and is in environmental psychology and earth and environmental science at the CUNY Graduate Center. He is the president of the board of the Participatory Budgeting Project and a member of the stakeholder advisory committee of the SRIJB. He has published work on technosci-ence, food sovereignty, and participatory budgeting and is finishing a book on participatory democracy. He received his PhD in philosophy (CUNY Graduate Center, 2002).

Walter Meyer

Walter Meyer, LEED-AP, is a principal at Local Office Landscape & Urban Design, LLC. Founded by Harvard Graduate School of Design classmates Meyer and Jennifer Bolstad, Local Office seeks to ameliorate the impact of cities on the sea, while protecting cities from sea surges. The firm has garnered accolades from across the disciplines of architec-ture, landscape architecture, public policy, science, and art.

Stephanie Miller

Stephanie Miller is currently a PhD candidate at Drexel University. She holds a BS in biol-ogy from Northeastern University (Boston, 2012) and an MS in environmental engineer-ing from Drexel University (Philadelphia, 2014). Her current areas of research include the evaluation of coastal green infrastructure during extreme events and agent modeling to study the long-term impacts of green infrastructure on urban resilience.

Franco A. Montalto

Franco A. Montalto, PE, PhD, is an associate professor in the Department of Civil, Archi-tectural, and Environmental Engineering at Drexel University, where he also directs the Sustainable Water Resource Engineering Laboratory. His expertise includes urban ecohy-drology, stormwater management, green infrastructure, hydraulic and hydrologic model-ing, and cross-cutting topics in urban sustainability, adaptation, and resilience planning. In addition to his academic teaching and research, he is the founder and president of eDesign Dynamics, LLC, an environmental consulting firm based in New York City. While this chapter was being drafted, he was also serving as the visiting scholar at the New York City Urban Field Station.

Philip Orton

Philip Orton is a research assistant professor at the Stevens Institute of Technology in Hoboken, New Jersey; a member of the New York City Panel on Climate Change; the NOAA-RISA funded Consortium for Climate Risk in the Urban Northeast (http://ccrun .org), and the executive council of the SRIJB. He has published more than twenty-five peer-reviewed articles, as well as three *New York Times* op-eds on climate change, coastal ecosystem health, flood protection, and coastal flooding. He holds a PhD in physical oceanography (Columbia University, 2010), and an MS in marine science (University of South Carolina, 1996).

Adam S. Parris

Adam Parris is the executive director of the SRIJB. He works on social and environmental change in U.S. coastal zones, Parris formerly served as the Climate Assessment and Services division chief and program manager for Regional Integrated Sciences and Assessments for the National Oceanic and Atmospheric Administration, and was a coastal planner for the San Francisco Bay Conservation and Development Commission. He is currently based at Brooklyn College, part of the City University of New York.

Laxmi Ramasubramanian

Laxmi Ramasubramanian, PhD, AICP, is an architect and urban planner. She is an associate professor in the Department of Urban Policy and Planning at Hunter College, part of the City University of New York (CUNY). She is the co-deputy director of the CUNY Institute for Sustainable Cities. Laxmi is deeply committed to participatory planning and community empowerment. Her first book, *Geographic Information Science and Public Participation,* was published by Springer-Verlag in 2010. She is currently working on a book on sustainable learning communities.

Erin Rieser

Erin Rieser is currently the volunteer manager for Habitat for Humanity of Dane County, Wisconsin. In addition to managing a program of 4,000 yearly volunteers, she works to improve community engagement programs and community services. She holds a bachelor's of architecture from Pratt Institute in 2006, and a master's in urban planning (2011) and a graduate certificate in GIS (2014) from CUNY Hunter College.

Hugh Roberts

Hugh Roberts is an associate vice president at ARCADIS, a leading global design and consultancy firm focused on natural and built assets. Trained at the University of Notre Dame, he is currently working on an integrated modeling program for Jamaica Bay.

Howard Rosenbaum

Dr. Howard Rosenbaum is director of the Wildlife Conservation Society's Ocean Giants Program. For more than twenty-five years, Dr. Rosenbaum's work has focused on innovative approaches for protecting marine species and their most biologically important habitats. Rosenbaum is also a senior scientist at the American Museum of Natural History and core affiliate faculty at Columbia University. He is a member of the U.S. delegation to the International Whaling Commission and the IUCN Cetacean Specialist Group, has been an associate editor for the journal *Marine Mammal Science,* holds the Conservation Seat for Stellwagen Bank National Marine Sanctuary's advisory council, and is a member of the Coastal and Ocean Advisory Panel for Monmouth and Rockefeller Universities. Rosenbaum received a BA from Hamilton College, was the recipient of a Thomas J. Watson Fellowship, and earned his PhD from Yale University.

Bernice Rosenzweig

Bernice Rosenzweig is a research associate at the Environmental Sciences Initiative of the CUNY Advanced Science Research Center. Her research focuses on urban ecological resilience from local to megaregion scales. Rosenzweig holds a PhD in environmental engineering (Princeton University, 2010).

Eric W. Sanderson

Eric W. Sanderson is a senior conservation ecologist at the Wildlife Conservation Society, serves on the executive council of the SRIJB and the board of the Natural Areas Conservancy in New York City, and teaches at New York University and Columbia University. He is the co-inventor of Visionmaker.nyc and the best-selling author of *Mannahatta: A Natural History of New York City* (Abrams, 2009). Other writings include *Terra Nova: The New World After Oil, Cars, and Suburbs* (Abrams, 2013), three edited volumes (including this one), and numerous peer-reviewed publications regarding wildlife and landscape conservation. Sanderson holds a PhD in ecosystem and landscape ecology (University of California, Davis, 1998).

Matthew D. Schlesinger

Matthew D. Schlesinger is the chief zoologist at the New York Natural Heritage Program, a program of the State University of New York College of Environmental Science and Forestry. He serves on the advisory board of the Natural Areas Conservancy. He has published scientific reports and peer-reviewed journal articles on biodiversity inventory and monitoring, habitat connectivity and climate change vulnerability modeling, at-risk tiger beetles, New York's dragonflies and damselflies, effects of urbanization on land birds, and leopard frogs of the eastern United States. Schlesinger holds a PhD in ecology (University of California, Davis, 2007).

William Solecki

William Solecki, MA, PhD, is a professor of geography at Hunter College, CUNY, and was the interim director of the SRIJB. His research interests include urban environmental change, urban spatial development, climate impacts, and adaptation. He has served on several U.S. National Research Council committees, including the Special Committee on Problems in the Environment. He is a founding member of both the Urban Climate Change Research Network and the International Human Dimensions Programme's Urbanization and Global Environmental Change Project. He was the former director of the CUNY Institute for Sustainable Cities and was a co-chair for Mayor Bloomberg's New York City Panel on Climate Change. Solecki has also contributed as a lead author to the IPCC Fifth Assessment Group II, Urban Areas Chapter.

Richard Stedman

Richard Stedman, PhD, is associate professor of Cornell University's Department of Natural Resources. Stedman's research focuses on the interaction between social and ecological systems. His training is in sociology, and he uses the theories and methodologies of this discipline as a lens for examining a broad array of human/environment conflicts. He is particularly interested in the challenges that rapid social and ecological changes pose for the sustainability of forested ecosystems, watersheds, and human communities.

R. Lawrence Swanson

R. Lawrence Swanson studies marine pollution in the coastal ocean. He contributed to policy debates about ocean dumping, marine debris, and sewage discharge criteria, and served as expert witness in marine boundary cases before the Supreme Court. He was a member of the Jamaica Bay Advisory Committee and is co-chair of the Long Island Sound Study Science/Technical Advisory Committee. He led several environmental programs when at the National Oceanic and Atmospheric Administration (NOAA) and was commanding officer on two NOAA research vessels. He holds a PhD in oceanography (Oregon State University, 1971) and was senior executive fellow at the Kennedy School, Harvard University (1983).

Keith G. Tidball

Keith G. Tidball is a senior extension associate in the Department of Natural Resources and director of the New York Extension Disaster Education Network, Cornell Cooperative Extension, Cornell University. He conducts research, extension, and outreach activities in the area of ecological dimensions of human security and is focused on natural resources management questions in places and time periods characterized by violence, conflict, disaster, or war. His writings include the edited volumes *Greening in the Red Zone: Disaster,*

Resilience and Community Greening, and *Expanding Peace Ecology;* the coauthored book *Civic Ecology: Adaptation and Transformation from the Ground Up;* and numerous journal articles. Tidball holds a PhD in natural resources (Cornell University, 2012).

John Waldman

John Waldman is an aquatic conservation biologist, with an emphasis on fishes. Waldman joined Queens College as a tenured professor of biology in 2004. For the previous twenty years he was employed by the Hudson River Foundation for Science and Environmental Research. He received his PhD in 1986 from the Joint Program in Evolutionary Biology between the American Museum of Natural History and the City University of New York, and an MS in marine and environmental sciences from Long Island University. Waldman has authored more than ninety scientific articles, edited a number of scientific volumes, and written several popular books, including *Heartbeats in the Muck: The History, Sea Life, and Environment of New York Harbor,* and, most recently, *Running Silver: Restoring Atlantic Rivers and Their Great Fish Migrations.* He also is an occasional essayist, including for the *New York Times* and *Yale Environment 360.*

Robert Wilson

Robert Wilson is an associate professor in the School of Marine and Atmospheric Sciences at Stony Brook University. His current research focuses on transport processes in estuaries. He holds a PhD from John Hopkins University (1974).

Christopher J. Zappa

Christopher J. Zappa is a Lamont Research Professor at the Lamont-Doherty Earth Observatory of Columbia University. Zappa is a leader in the field of air-sea interaction with extensive in situ and airborne observational-based expertise. He is dedicated to understanding the processes that affect ocean–atmosphere interaction and their boundary layers. His focus includes wave dynamics and wave breaking, upper-ocean processes, polar ocean processes, and coastal and estuarine dynamics. He has led a continuing evolution of the development of measurement systems. Zappa is a member of the University-National Oceanographic Laboratory System (UNOLS) Science Committee on Oceanographic Aircraft Research and the NASA Sea Surface Temperature Science Team. He has published more than forty refereed papers. Zappa holds a PhD in ocean and applied physics (University of Washington, 1999).

Chester B. Zarnoch

Chester B. Zarnoch is an associate professor of environmental studies and biology at Baruch College, City University of New York (CUNY), and is graduate faculty in the biology program at CUNY's Graduate Center. He has been an active researcher in Jamaica Bay since 2001 and has published several papers on his work with a focus on shellfish biology and sediment nitrogen cycling. His current research aims to describe the biological and physical processes that influence ecosystem services derived from restored habitats in eutrophic estuaries. Zarnoch holds a PhD in biology (Graduate Center, CUNY, 2006).

Index